SMOKE AND MIRRORS

How to bend facts and figures to your advantage

Nicholas Strange

D1419116

A & C BLACK · LONDON

To JPAS, that le

First published in 2007 by
A & C Black Publishers Ltd
38 Soho Square, London W1D 3HB

British Library Cataloguing in Publication Data
A CIP record for this book is available from the
British Library.

ISBN-10: 0-7136-7924-7
ISBN-13: 978-0-7136-7924-3

A & C Black uses paper produced with elemental
chlorine-free pulp, harvested from managed
sustainable forests.

Designed by Fiona Pike, Pike Design, Winchester
Typeset by Margaret Brain
Printed in the United Kingdom by MPG Books

CONTENTS

1 INTRODUCTION

This book is about turning numbers into effective graphics, to persuade, impress or confuse. It explains the graphic techniques available to make the best of a good, mediocre or non-existent case without, quite, lying outright.

We need such techniques because figures do not speak for themselves. Numbers alone seldom make a convincing case or polish their author's image – the twin goals of that other great mind bender, rhetoric. While rhetoric deals in qualitative argument, its quantitative equivalent is graphics. As rhetoric has declined in popularity, so graphics have risen – along with our acceptance of quantitative arguments. In graphics, figures finally find their own means of expression.

The computer revolution and powerful software mean we can now turn numbers into charts on every desktop without acquiring either graphical or numerical skills. As a result, accident has overtaken design as the main source of graphic deceit. This book aims to straighten out this tangled web and to ensure that graphic deceit – when it occurs – is deliberate, sufficiently distorting to be worthwhile, subtly difficult to detect and, by combining distortion and subtlety, successful.

But this book is not about which computer key to push or icon to click. If you need a handbook to use modern software there's something badly, probably incurably, wrong either with you or the software. Yet most bookshops usually stock a dozen or more handbooks at least touching on the visual display of data. Nearly all are about how to get the software to turn numbers into lines and colours rather than about designing the finished product – the chart or graph – itself. In this respect most Excel or PowerPoint handbooks are like a nightmare driving school that teaches all there is to know about the engine and transmission but fails to recommend stopping at red lights.

This book also tries to give credit where credit is due by displaying and attributing real, published examples of the successful application of the principles it preaches. However the technical and organisational complexity of the publishing process (traditional or electronic) obscures the motivation of the perpetrators. So a reasonable doubt usually persists about whether we are observing the fruits of criminal genius or just honest incompetence. Nothing in this book should be interpreted as favouring either of these hypotheses over the other. After all, it is precisely this ambiguity that separates the truly great graphical deception from the merely workmanlike.

What distinguishes data tables from graphics is explicit comparison and the data selection that this requires. While a data table obviously also selects information, this selection is less focused than a chart's on a particular comparison. To the extent that some figures in a table are visually emphasised, say in colour or size and style of print, the table is well on its way to becoming a chart. If you're making no comparisons – because you have no particular message and so need no selection (in other words, if you are simply providing a database, number quarry or recycling facility) – tables are easier to use than charts.

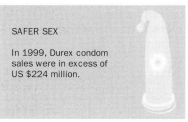

SAFER SEX

In 1999, Durex condom sales were in excess of US $224 million.

Mackay, *Atlas of Human Sexual Behavior,* 2000

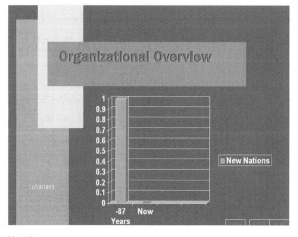

Abraham Lincoln: 'Fourscore and seven years ago our fathers brought forth on this continent a new nation, conceived in liberty and dedicated to the proposition that all men are created equal...'

Norvig

As a data source, a place for quarrying or simply retrieving numbers, the best graphic is never going to be better than a standard data table. Attempting to provide a database as well as support a message obscures the message as much as it impedes data retrieval. So, if in honest doubt, go for the table. If dishonest, go for the chart.

In the extreme case of a single number ('sales were in excess of US$224 million'; 'Fourscore and seven years ago, our fathers brought forth...'; 'into the valley of death rode the six hundred') the lack of any comparison makes graphic treatment unnecessary, though the attempt to turn single numbers into 'charts' can be quite diverting in its lunacy and even instructive in prompting us to think about why we use graphics at all.

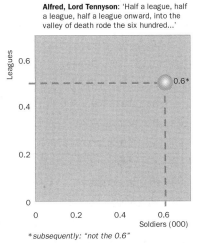

Alfred, Lord Tennyson: 'Half a league, half a league, half a league onward, into the valley of death rode the six hundred...'

*subsequently: "not the 0.6"

Types of distortion

Types of graphic deceit abound, but from the point of view of the victim – the presentation audience, the reader of non-fiction, the fellow scientist, the actual or potential client – it always comes down to the use of tricks to bend the relationship between the graphic impression and the underlying data.

There are three groups of deceitful techniques, defined according to how and where they gain their leverage and the graphic elements to which they are best applied.

1 Distorting values, usually the vertical scale or y-axis, by manipulating either the data itself, for example by using unexpected or inappropriate unit definitions, or by bending the graphical representation of the data, for example by hiding the zero point or by logarithmic spacing of a scale. These techniques make the values of the data points – and the differences between them – look bigger or smaller than they actually are.

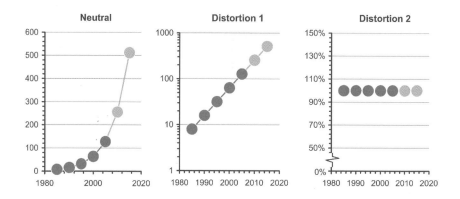

2 Distorting category (including time), usually the horizontal scale or x-axis, for example by comparing categories that are incomparable since not mutually exclusive or by using an irregular time scale. As with the value axis, preliminary distortion of the data is harder to detect than later graphical tricks.

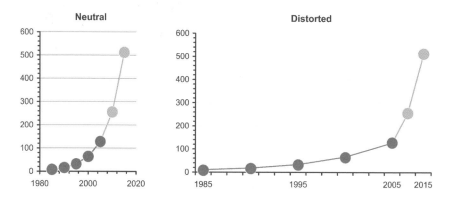

3 Distorting the whole chart or indeed the relationship between the chart and its message. These techniques include ways of enhancing the plausibility of a basically dodgy graphic, like over-footnoting, and some common logical fallacies and mathematical misunderstandings. Apart from distortion techniques, they also include ways of making the data inaccessible, and hence the differences among data points less striking, for example by hiding it behind chart junk like company icons, pointing gauntlets, and absurdly over-sized arrows and star bursts or by using illegible text and confusing text orientations.

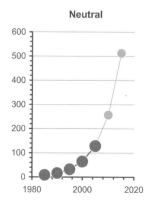

Neutral

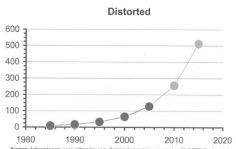

Distorted

Sources: Adherents.com, www.adherents. com, Religion by Location, accessed on 21.3.2004.Eurobarometer, Standard Euro-barometer 42, sections 10.5 & 10.6 1994, accessed on 21.3.2004. Institut CSA, Les Francais et leurs Croyances, Groupe CSA-TMO, Mars 2003. Jehovah's Witnesses Official Web Site, www.watchtower. org/statistics/worldwide_report, accessed on 21.3.2004 (Jehova's Witnesses only). REMID, Religionswissen-schaftlicher Medien- und Informationsdienst e.V., www.uni-leipzig.de, accessed on 21.3.2004. UK Statistics, Social Trends 30, Church Membership, 1970 - 1990, ST301319 and Census 2001, Belonging to a Religion, table 13.24. University of Michigan, World Values Surveys 1990, 1991 & 1995 -7, news release 10.12.1997, www.umich.edu, accessed on 22.3.2004

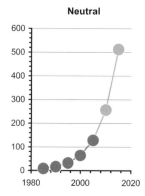

Neutral

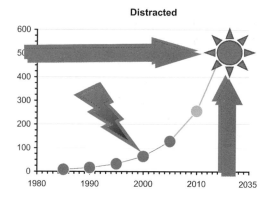

Distracted

The full galaxy of deception techniques is impressively large and can almost be guaranteed to provide a technique for every occasion.

So this book assumes, when first you practise to deceive, that you have already taken a view (as they say in Whitehall) about whether you want to bend the values, the categories or the relationship between them, or simply to confuse or even repel the audience. The detailed descriptions of the techniques are also arranged in these three groups in this book, as the starting point in practice will usually be an existing 'neutral' chart, on which leverage points can be selected and techniques applied at leisure.

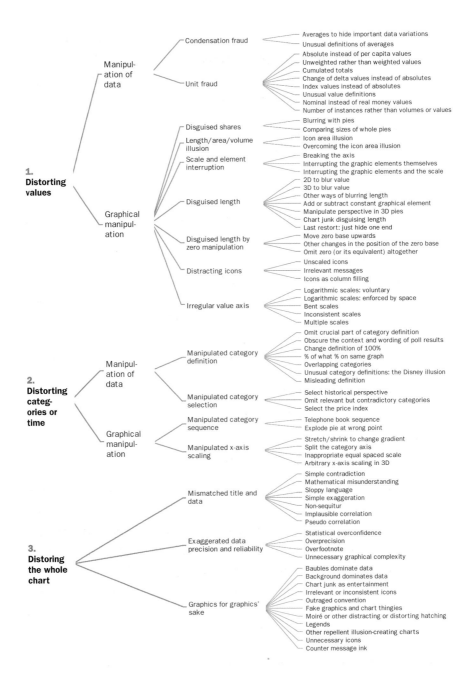

1.
Distorting values

Manipul-ation of data
- Condensation fraud
  - Averages to hide important data variations
  - Unusual definitions of averages
- Unit fraud
  - Absolute instead of per capita values
  - Unweighted rather than weighted values
  - Cumulated totals
  - Change of delta values instead of absolutes
  - Index values instead of absolutes
  - Unusual value definitions
  - Nominal instead of real money values
  - Number of instances rather than volumes or values

Graphical manipul-ation
- Disguised shares
  - Blurring with pies
  - Comparing sizes of whole pies
- Length/area/volume illusion
  - Icon area illusion
  - Overcoming the icon area illusion
- Scale and element interruption
  - Breaking the axis
  - Interrupting the graphic elements themselves
  - Interrupting the graphic elements and the scale
- Disguised length
  - 2D to blur value
  - 3D to blur value
  - Other ways of blurring length
  - Add or subtract constant graphical element
  - Manipulate perspective in 3D pies
  - Chart junk disguising length
  - Last restort: just hide one end
- Disguised length by zero manipulation
  - Move zero base upwards
  - Other changes in the position of the zero base
  - Omit zero (or its equivalent) altogether
- Distracting icons
  - Unscaled icons
  - Irrelevant messages
  - Icons as column filling
- Irregular value axis
  - Logarithmic scales: voluntary
  - Logarithmic scales: enforced by space
  - Bent scales
  - Inconsistent scales
  - Multiple scales

2.
Distorting categ-ories or time

Manipul-ation of data
- Manipulated category definition
  - Omit crucial part of category definition
  - Obscure the context and wording of poll results
  - Change definition of 100%
  - % of what % on same graph
  - Overlapping categories
  - Unusual category definitions: the Disney illusion
  - Misleading definition
- Manipulated category selection
  - Select historical perspective
  - Omit relevant but contradictory categories
  - Select the price index

Graphical manipul-ation
- Manipulated category sequence
  - Telephone book sequence
  - Explode pie at wrong point
- Manipulated x-axis scaling
  - Stretch/shrink to change gradient
  - Split the category axis
  - Inappropriate equal spaced scale
  - Arbitrary x-axis scaling in 3D

3.
Distoring the whole chart

- Mismatched title and data
  - Simple contradiction
  - Mathematical misunderstanding
  - Sloppy language
  - Simple exaggeration
  - Non-sequitur
  - Implausible correlation
  - Pseudo correlation
- Exaggerated data precision and reliability
  - Statistical overconfidence
  - Overprecision
  - Overfootnote
  - Unnecessary graphical complexity
- Graphics for graphics' sake
  - Baubles dominate data
  - Background dominates data
  - Chart junk as entertainment
  - Irrelevant or inconsistent icons
  - Outraged convention
  - Fake graphics and chart thingies
  - Moiré or other distracting or distorting hatching
  - Legends
  - Other repellent illusion-creating charts
  - Unnecessary icons
  - Counter message ink

Choosing to deceive

When choosing a technique of deception to suit your own particular case you need to bear two, often contradictory, considerations in mind.

First, how powerful is the technique? On the assumption that outright lying is not an option, the most serious constraint is the underlying data.

PDQ (POTENTIAL DECEIT QUOTIENT)

Ratio of perceived value in the chart to original data value.
A gross value that can vary from as little as 0.001:1 to 1,000:1.

Other limits are set by the medium (graphics in printed books, for example, are inherently harder to manipulate than overhead slides), the acuity of your audience and even the size of the chart you can use. But some techniques are nearly always more powerful than others and this is reflected in this book by an estimate of each technique's PDQ (potential deceit quotient): the ratio of perceived value in the finished graphic to the original value in the data. Techniques with lower PDQs are not necessarily to be despised. You may, after all, only need a small adjustment and the lower the PDQ, the easier it is to explain away later.

STD (SORE THUMB DISCOUNT)

Proportion of the deceit quotient that gets noticed, or at least unconsciously, discounted, by the viewer of the chart.
0.0 = unnoticed, not discounted at all
1.0 = completely discounted, because very obviously bent.

Which brings us to the second, equally important, consideration in choosing a deceit technique: how likely is it to be noticed? Will it stick out like a sore thumb or will it go unnoticed or at least be less than fully discounted, consciously or unconsciously, by the viewer? This is expressed as the STD (sore thumb discount). An STD of 0 means that the manipulation is all but invisible under normal circumstances; of 1 that it sticks out like a whore at a christening and makes the use of the technique almost entirely pointless.

PDQ x (1 – STD) = net (i.e. discounted) expected value of the technique

Multiplying PDQ and STD gives a good measure of the real value of the technique, taking account of both its power and its subtlety.

Most books on visualising quantitative data start with a chapter on 'choosin' and usin'' chart types appropriate to the data to be displayed. Even if you have deceit in mind, this is still an important consideration because the PDQ of a technique obviously depends partly on the power of the chart type itself. While there are occasions when the use of a spectacularly inappropriate type can lay a useful smoke screen, behind which you can withdraw gracefully to higher ground, it is generally a good idea to be conservative and conventional in choosing what sort of chart to use.

Luckily, the conventional rules for choosing types of chart usually leave quite a lot of room for personal taste. There's not all that much difference in most situations between the suitability of a column or bar chart, an x/y line diagram or even a scatter plot. The crucial consideration is which type of chart leaves the most room for graphic deceit. So the following survey of the nine most commonly used chart types looks not only at what information each type shows best but also at the individual graphic elements used to show it and their vulnerability to manipulation.

Pies and **donuts** are considered best at showing shares. The way they do this is not quite as simple as it first appears. For shares between one quarter and three-quarters, pies rely on our strong ability to identify straight lines (halves) and right angles (quarters). We are not as good at more subtle gradations, though as we see below, one of the few characteristics that Mercedes owners and members of CND have in common is probably a highly developed sense of thirds.

The donut, its spelling betrays its origins, is nearly always more deceit friendly than the pie, despite being modelled on a life-saving ring.

Pies and **donuts**

show (in order of effectiveness):
shares
absolute values
differences
ranks
using (in order of visual power):
straightness of lines
angles around centre
slice circumference
areas
Distortion exploits:
donut to hide angles and straightness of line
perspective to change angles and slice
    circumferences
third dimension to manipulate areas of slices

This is because the hole destroys the second most important value-defining element, by hiding the slice angles in the middle.

Pies are particularly vulnerable to 3D perspective, which can change both the angles and the areas of slices by at least a factor of two. Even Mercedes bends under such treatment.

The fundamental vulnerability of the donut form also underlines how difficult it is to compare non-adjacent angles. For blurring changes in share, usually over time, there's nothing quite like a pair of pies, if possible of different size, whose slices are arranged in different orders. The last is important, as we are quite good at seeing whether lines are parallel or not and can use this ability to compare slices from different pies as long as one side of each slice is at the same angle to the vertical.

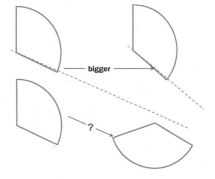

We tend to think of a series of different sized circles as just a group of pie charts, but this is only usefully true if each circle is segmented. Take the two-dimensional arrays purporting to sort a company's product market segments into stars for further investment, dogs to be sold, cows to be milked of further investment funds and don't knows to be analysed further at huge expense by the consultant making the presentation. These 'portfolio matrices' represent the market size of each segment by the size of the circle that also shows its strategic potential in terms of market attractiveness and the strength of the company's market position.

Such **multiple circles** open the door on a wonderland of deceit based on the ambiguity of whether data values are proportional to the radius (r) of the circles or their area (πr^2) or even their virtual volume ($4\pi r^3/3$). The same considerations apply if you use single icons rather than circles or balls: for example, grim-looking doctors to denote NHS personnel totals, or rockets to show the size of nuclear arsenals. Is it the height

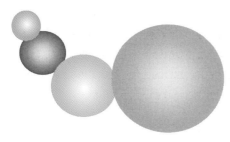

Multiple pies or **circles**
show (in order of effectiveness):
absolute values
differences
ranks
using (in order of visual power):
areas
diameters
Distortion exploits:
ambiguity between areas and diameters (m^2
   problem)
distance between pies to be compared (angles etc.)

of the doctor icon, its waist measurement, the area it subtends on the retina or its apparent volume that corresponds to that year's staff total? There's no knowing, and PDQs can rise to double figures before thumbs start to get uncomfortably sore. Multiple pies, in which the circles are also segmented to show shares that need a ruler, protractor and calculator to compare, are even better.

There are some chart types that occasionally appear in print but are so bad that they serve neither honesty nor deceit. Among these monuments to human ingenuity at the expense of common sense are the concentric donut and overlapping segments. The concentric donut is really just a bar or column chart bent back on itself to save space. However as anyone who has ever watched a two or four hundred metre race will know, to make sense of the order of arrival at the tape you have to stagger the start to take account of the bend in the track. Blithely ignoring this problem, the concentric donut uses π to diminish the difference between the inner and the outer absolute values by anything up to 2½ times.

Overlapping segment charts overemphasise the smallest value, the only one shown in its entirety, and so can occasionally have a special

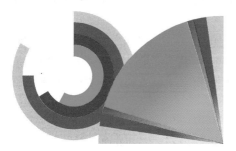

Concentric donuts and **overlapping
segments**
show (in order of effectiveness):
absolute values
ranks
differences
using (in order of visual power);
slice circumference
areas
Distortion exploits:
perspective to change angles and slice
   circumferences
third dimension to manipulate areas of slices

role in drawing attention away from a low value or poor performance. Unfortunately, both these forms are so unusual that they stand out a mile. So STDs are very large and after the discount these chart types probably aren't worth the effort, especially as few business graphics applications are capable of constructing them automatically.

Bars and columns are the workhorses of quantitative visualisation and hence the Dome of the Rock, the St Peter's – or at least the Wailing Wall – of graphic deceit. They are usually recommended for comparisons among categories or over time. They can be blurred by almost any 3D treatment and their perceived lengths manipulated either in the scale, in the bar or column itself, or more trickily by the use of an optically distorting background.

They have a delightful built-in ambiguity about whether we should look at their length, area

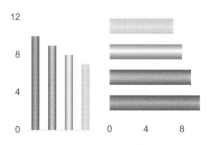

Bars and **columns**

show (in order of effectiveness):
absolute values
differences
ranks
using (in order of visual power):
lengths
areas
Distortion exploits:
condensation or unit fraud
length/area/volume ambiguity
irregular/broken/multiple scales
inappropriate categories
camouflaged length
displaced zeroes
distracting chart junk and many more

or volume. Habits have changed in the last hundred years. At the beginning of the twentieth century, the default expectation was that a bar chart represented values proportional to the area of the bars, which tended to vary quite widely in width. The modern expectation, shown here, is of equal width bars in which the value is thus proportional to the height and hence, of course, to the area too. This avoids the area illusion but falls prey to another.

There is a general feeling, none the less powerful for being unspoken, that all bars in a chart are in some sense equally important, however much they may differ in height or length. This quite unjustified default position then leads to absurdities like the EU health statistics, in which the figures for infant deaths per thousand live births (so called infant mortality) in, say, Luxemburg, are given equal billing to those in the UK with a hundred times the population. We need a Martin Luther to nail this graphical bull by insisting that bar and column widths should be made proportional to the category's importance whenever the value displayed is not an absolute number. Nobler considerations aside, the opportunities for deceit would double overnight as we exploited all the possible meanings of 'important'.

Scatter plots are recommended for showing, even investigating, correlations. With or without trend or correlation lines, they tend to frighten non-academic audiences. Although their scales are open to exactly the same manipulations as those of bar and column charts, they have a higher plausibility rating. Management consultants anxious not to damage their street cred with a very moderately educated salariat tend to shun them. To politicians, who need a stiff drink before saying 'percent' rather than the homelier 'pence in the pound', scatter plots are about as welcome in their reading lists as a former colleague's diaries. Scatter plots are, quite understandably, much used by social scientists keen to deduce a cause from a correlation.

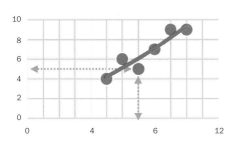

Scatter diagrams with **correlation** or **trend lines**

show (in order of effectiveness):
two absolute value series
differences in two dimensions
correlation or groups
using (in order of visual power):
scale (implicit) length
grouping
angles of trend lines
Distortion exploits:
condensation or unit fraud
irregular/broken/multiple scales
displaced zeroes
suggested trends distracting chart junk and many more

Their ability to display multivariate data (two or more characteristics of the sample population at the same time) give scatter plots an analytical power unmatched by simpler, univariate forms such as pie charts. So if you can find an audience that won't rush screaming to the exit at the mere sight of two axes on one chart, if neither represents time nor a category, scatter plots offer fertile ground for manipulation.

This also goes for the trend lines that are often superimposed on the data points. Most applications packages offer several trend-fitting formulae at the push of a button. Remember that all these packages will fit some sort of trend line to any data. They would unblushingly put a trend line through the random numbers used at GCHQ as the base for uncrackable codes. So it's worth experimentally hitting the trend button whenever you finalise a scatter plot; it's a bit like looking in the coin return of parking ticket machines for forgotten change but with potentially higher rewards. The chances of a non-academic audience asking about the quality of the fit, let alone understanding your answer (just say 'we used a standard chi square on this one' before turning back to the screen with a deprecating shrug), are negligible.

X/y line diagrams are best restricted, in the view of most commentators, to comparisons over time. They are the other popular icon of quantitative visualisation after bars and columns, whether clipped to the foot of an NHS bed to record axillary temperature or from the FTSE report to show share price. They are really just scatter plots in which the independent variable is time or, looking at it the other way, just column charts in which the columns have been replaced by dots and lines.

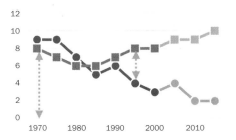

X/y line diagrams

show (in order of effectiveness):
absolute values
change over time
differences
ranks
using (in order of visual power):
scale (implicity) length
angles (if dots connected)
Distortion exploits:
condensation or unit fraud
irregular/broken/multiple scales
displaced zeros
distracting chart junk and many more

They are particularly vulnerable to uneven stretching of the time axis, which can change the slope of the curve dramatically without resort to the higher mysteries such as logarithmic value scales. In real life this distortion often occurs as a by-product of entirely innocent attempts to use all available information in the chart: the original technicolor data yawn. As statistics are collected more frequently nowadays, the more modern the data, the shorter the time periods tend to get. So even if the deceit is noticed, it will usually be put down to over-enthusiasm in presenting full information or forgiven as an understandable concession to lack of space on the page. On the value axis, x/y line diagrams are open to the same subversion as scatter plots.

X/y area diagrams seem at first glance just to be line diagrams with extra shading, but in fact they offer some useful additional manipulation opportunities.

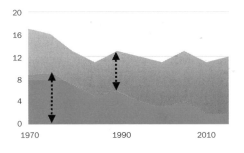

X/y area diagrams

show (in order of effectiveness):
absolute values
change over time
differences
ranks
shares of a total (if stacked)
using (in order of visual power):
scale (implicity) length
angles (if dots connected)
Distortion exploits:
condensation or unit fraud
irregular/broken/multiple scales
displaced zeroes
distracting chart junk and many more

The data categories in x/y area diagrams used always to be stacked rather than zero based. The top line of the top area thus gave the total of all data points for the period. In situations in which the total was important and 'stacking' was thus required, it was easier to stack areas than lines. The other practical consideration was simply that data points that were lower in value than those in front of them disappeared if the areas were not stacked. Increasing use of 3D area diagrams, in which you can at least see the trench into which the invisible, unstacked data has fallen, seems to have made designers careless of losing data. Nowadays it is often unclear if the areas are stacked or all are based on zero.

If one of your categories has developed in the period covered by the chart in ways that you would prefer went unnoticed, put it at the top and other categories, the more erratic the better, below it. It will be almost impossible to read the values for categories at the top and, if taxed with conspiring to hide bad news, you can always say that you did put the figures in question in the most noticeable position at the top of the chart.

Shading can be used to dramatise a 'Himalayan' profile of the top line and, more importantly, to obscure the grid lines leading to data we want to keep vague. Unstacked areas can be 'lost' altogether.

Divided bars or **columns** are for comparisons over time or between shares and offer all the opportunities of their undivided cousins for scale manipulation and area illusions. As they are much used for showing differences over time, they are particularly vulnerable to perspective effects in which the apparently more distant columns are difficult to align with the scale.

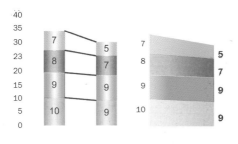

Divided bars or **columns**

show (in order of effectiveness):
differences
absolute values
shares of a total
using (in order of visual power):
lengths
angle of (connecting) lines
areas
Distortion exploits:
condensation or unit fraud
length/area/volume ambiguity
irregular/broken/multiple scales
inappropriate categories
camouflaged length
displaced zeroes
distracting chart junk and many more

Radar charts are almost always the result either of space-saving attempts or of doubtful theories about the desirability of 'symmetrical' plots, in which scores on all dimensions are similar, so giving an approximation to a circle. Their scales offer unlimited scope for manipulation in achieving this lunatic ambition. The Nightingale plot on the right is seldom seen. It is as difficult to draw as to read – and consequently easy to distort – but its unfamiliarity imposes an STD close to zero. On the other hand, it was the chart type showing the deaths in the Scutari hospital before and after her improvements, that Florence Nightingale used to persuade the British government to reform military medicine. Victorian politicians were more tolerant of intellectual effort, whether in themselves or others.

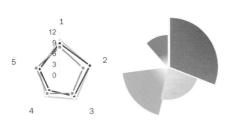

Multi-axis charts

show (in order of effectiveness):
differences
absolute values
ranks
using (in order of visual power):
Radar: distance from centre, roundness, symmetry
Nightingale: area or slice angle, area or
circumference
Distortion exploits:
preference for symmetry
scale manipulation

Before looking at the individual techniques, it's worth at least glancing at the question of the power of graphics in general. So the next chapter gives examples of how graphic failure seems to have been a vital link in two famous, modern disasters and in the avoidance of a much older one.

If you are already convinced of the power of graphics, skip directly to Chapter 3, which plunges enthusiastically into the distortion of values.

2 THE POWER OF GRAPHICS

Although most graphical manipulation just distorts our understanding of the world and affects only our careers, incomes, job satisfaction, sex life, children's future and whether we have a realistic chance of getting a Nobel prize, some really matters.

This chapter looks at three examples, spread over 150 years, of graphics or semi-graphics that failed ignominiously to communicate with their chosen audiences, in spite of the fact that we now know that their intended messages were not only too true but also a matter of life and death. The two first examples are both drawn from the American space-shuttle programme, not because it is unusually incompetent, rather the reverse, but because the official enquiries into the Challenger and Columbia disasters gave the general public an unusually deep insight into the mistaken decisions that were taken and the materials, particularly graphically treated materials, on which they were based. The third example is from the mid-nineteenth century and looks at the graphics used by Dr John Snow to convince the scientific community of the true causes of the Soho cholera outbreak – at which he failed – and persuade the authorities to take the necessary preventative measures – at which he surprisingly succeeded.

For our purposes, both NASA space-shuttle examples demonstrate how communications failure can already arise from mistakes in the pre-graphical phases of data preparation and how this failure can be camouflaged and even reinforced by purely graphical techniques in its final presentation. In the Challenger case, even the graphics in the report of the Presidential Commission set up to investigate the tragedy failed to break entirely free from the original presentational distortion.

Challenger: fatal category choice and sequence

For the two minutes after launch, more than 80% of the total thrust of the space-shuttle is provided by two gigantic solid-fuel booster rockets on either side of the orbiter. The boosters, despite containing over 450 tonnes of solid fuel each, still share two important characteristics with their Guy Fawkes ancestors. Once the blue touch paper has been lit, neither can be turned off and both must contain and direct very hot gases under high pressure towards the hole at the back. But while the Guy Fawkes rocket's gunpowder is contained in a seamless tube of rolled paper, the boosters' powdered aluminium/ammonium perchlorate (basically the traditional terrorist's weedkiller recipe) is contained in three tubular sections bolted together end to end.

Anyone whose car engine has experienced valve or cylinder head failure can testify to the astonishing speed with which the smallest leak of combustion gases can erode its way to total failure in seconds, even at the moderate temperatures in an ordinary car. So at the booster's vastly higher temperatures, it is crucial to prevent the slightest gas leak (known to NASA as 'blow-by'), let alone actual erosion, where the sections are joined. Two enormous O-shaped rubber washers slotted into matching grooves at each joint complete a gas-tight seal. Unfortunately, rubber, be it ne'er so hi-tech, loses elasticity and therefore some of its sealing qualities, at low temperatures.

Challenger's tenth flight (STS-51-L) had been scheduled for 22 January 1986, but there were several technical problems and by the evening of 27 January, the shuttle was still on the ground waiting for launch the next morning at 11:38 local time. NASA, knowing that there had been blow-by and even erosion of the O-rings on several previous flights, was worried that the forecasted temperature of 26°–29°F (–2° to –3°C), which was much lower than that at any previous launch, would reduce the resilience of the O-rings, leading to possibly catastrophic erosion at the joints of the booster segments. NASA asked Morton-Thiokol, the manufacturer of the booster, to recommend for or against launch the next day. That evening Morton-Thiokol faxed a 13-page presentation to NASA arguing against launch on the grounds of dangerously low ground temperatures. But despite having asked the pertinent question, NASA was now unconvinced of the dangers and

Morton-Thiokol management changed their advice in a teleconference that same evening. The shuttle was launched on time next morning and exploded 70 seconds after lift-off, when the flames escaping from one of the booster-joints ignited the main fuel tank. The analytical and presentational failure of those 13 pages to convince first Morton-Thiokol and then NASA management of the hazards of launching at un-Florida-like low temperatures was what opened the door to disaster. So why did the presentation fail to persuade NASA to postpone the launch?

Of the 13 pages the first is the title page, which reads limply but reasonably enough:

TEMPERATURE CONCERN ON
SRM JOINTS
27 JAN 1986

(SRM being the abbreviation for Solid Rocket Motor). But what do you do if you have a 'concern'? A natural human reaction is to focus on the instances that have given concern – in this case, six of seven flights showing major and minor O-ring malfunction between November 1981 and the last flight before Challenger on 12 January 1986, listed on page 2 of the report. Seventeen other flights without O-ring problems are ignored throughout. Pages 3 to 5 and 7 to 10 make the point that the secondary O-ring is of limited value if the primary permits erosion, and

HISTORY OF O-RING DAMAGE ON SRM FIELD JOINTS

	SRM No.	Cross Sectional View			Top View		Clocking Location (deg)
		Erosion Depth (in.)	Perimeter Affected (deg)	Nominal Dia. (in.)	Length Of Max Erosion (in.)	Total Heat Affected Length (in.)	
61A LH Center Field**	22A	None	None	0.280	None	None	36°--66°
61A LH CENTER FIELD**	22A	NONE	NONE	0.280	NONE	NONE	338°-18°
51C LH Forward Field**	15A	0.010	154.0	0.280	4.25	5.25	163
51C RH Center Field (prim)***	15B	0.038	130.0	0.280	12.50	58.75	354
51C RH Center Field (sec)***	15B	None	45.0	0.280	None	29.50	354
41D RH Forward Field	13B	0.028	110.0	0.280	3.00	None	275
41C LH Aft Field*	11A	None	None	0.280	None	None	--
41B LH Forward Field	10A	0.040	217.0	0.280	3.00	14.50	351
STS-2 RH Aft Field	2B	0.053	116.0	0.280	--	--	90

*Hot gas path detected in putty. Indication of heat on O-ring, but no damage.
**Soot behind primary O-ring.
***Soot behind primary O-ring, heat affected secondary O-ring.

Clocking location of leak check port - 0 deg.

OTHER SRM-15 FIELD JOINTS HAD NO BLOWHOLES IN PUTTY AND NO SOOT NEAR OR BEYOND THE PRIMARY O-RING.

SRM-22 FORWARD FIELD JOINT HAD PUTTY PATH TO PRIMARY O-RING, BUT NO O-RING EROSION AND NO SOOT BLOWBY. OTHER SRM-22 FIELD JOINTS HAD NO BLOWHOLES IN PUTTY.

Report of Presidential Commission on Space Shuttle Challenger Accident, 1986

HISTORY OF O-RING TEMPERATURES
(DEGREES - F)

MOTOR	MBT	AMB	O-RING	WIND
Dm-1	68	36	47	10 mPH
Dm-2	76	45	52	10 mPH
Qm-3	72.5	40	48	10 mPH
Qm-4	76	48	51	10 mPH
SRM-15	52	64	53	10 mPH
SRM-22	77	78	75	10 mPH
SRM-25	55	26	29	10 mPH
			27	25 mPH

1-D THERMAL ANALYSIS

Report of Presidential Commission on Space Shuttle Challenger Accident, 1986

explain the expected link between lower launch temperature and erosion with the help of bench test data. Pages 12 and 13 present a rather muffled conclusion. Of the four pages containing data on actual missions – three (pages 2, 6 & 10) record O-ring damage and one (page 11) records launch temperature. There is no page containing both temperature and damage data for any flight, the subject of NASA's enquiry, let alone for all the 24 flights before Challenger.

It is difficult to compare this table of O-ring temperatures (page 11) with the chart showing blow-by history (page 6) as both contain large amounts of irrelevant data. The first four motors on the temperature charts are largely irrelevant horizontal test firings of development and

BLOW-BY HISTORY
SRM-15 WORST BLOW-BY
o 2 CASE JOINTS (80°), (110°) ARC
o MUCH WORSE VISUALLY THAN SRM-22

SRM 22 BLOW-BY
o 2 CASE JOINTS (30-40°)

SRM-13A, 15, 16A, 18, 23A 24A
o NOZZLE BLOW-BY

Report of Presidential Commission on Space Shuttle Challenger Accident, 1986

quality assurance boosters, and the last is the not yet launched Challenger shuttle itself. The last two lines on page 6 concern nozzle blow-by, a different problem. Nonetheless it was the comparison of these two pages that finally torpedoed the Morton-Thiokol engineers' initial and correct assessment, that launches at less than 53°F would be extremely hazardous.

RECOMMENDATIONS :

° O-RING TEMP MUST BE $\geq$ 53 °F AT LAUNCH

DEVELOPMENT MOTORS AT 47° To 52°F WITH
PUTTY PACKING HAD NO BLOW-BY
SRM 15 (THE BEST SIMULATION) WORKED AT 53 °F

° PROJECT AMBIENT CONDITIONS (TEMP & WIND)
To DETERMINE LAUNCH TIME

Report of Presidential Commission on Space Shuttle Challenger Accident, 1986

It was pointed out during the teleconference on the evening before launch, that the coldest launch (SRM-15 at 53°F) and the warmest launch (SRM-22 at 75°F) had both shown signs of O-ring damage. In other words, something other than launch temperature was causing O-ring damage and the forecasted 27°–29°F for the morrow was of no consequence. Relying on an inadequate, indeed hopelessly rickety, sample size of two flights (the only ones for which they had supplied both damage and temperature data) the Morton-Thiokol engineers were blown forensically and rhetorically out of the water by their own and NASA's managers. The original conclusion (page 13), despite only giving details that tended to soften the message, had still been clear enough, stating 'O-RING TEMP MUST BE $\geq$ 53°F AT LAUNCH'. It was withdrawn.

It's enlightening to recall the data actually available to Morton-Thiokol (and hence potentially to NASA) that evening, but not fully used. The table is constructed from Flight Readiness Reviews normally carried out ten days before each launch, which included data from previous launches, and shows two things with embarrassing clarity.

First, and most importantly, when the data for all flights and incidents to field joints is displayed in order of temperature, the relationship between low launch temperature and subsequent O-ring distress leaps

Temperature (C°) of the field Joint at launch No.	Shuttle Flight (STS) No.	SRM* No.	Date	Damage to field joints	
				Erosion of O-ring	Blow-by
*-3 to -2** *	*51-L*	*SRM 25*	*28.01.86*	*waiting for launch clearance*	
12	51-C	SRM 15	24.01.85	3	2
14	41-B	SRM 10	03.02.84	1	
14	61-C		12.01.86	1	
17	41-C	SRM 11	06.04.84	1	
19	1		12.04.81		
19	6		04.04.83		
19	51-A		08.11.84		
19	51-D		12.04.85		
20	5		11.11.82		
21	3		22.03.82		
21	2		12.11.81	1	
21	9		28.11.83		
21	41-D	SRM 13	30.08.84	1	
21	51-G		17.06.85		
22	7		18.06.83		
23	8		30.08.83		
24	51-B		29.04.85		
24	61-A	SRM 22	30.1.0.85		2
24	51-I		27.08.85		
24	61-B		26.11.85		
26	41-G		05.10.84		
26	51-J		03.10.85		
27	4		27.06.82	?	?
27	51-F		29.07.85		

Facts mentioned in report to NASA on 27.01.1986 in red
*SRM = Solid Rocket Motor (only for launches mentioned in report)
**forecast

Data from: *Report of Presidential Commission on Space Shuttle Challenger Accident,* 1986

to the eye. No launch above 24°C displayed O-ring distress; all launches below 19°C did. Second, the Morton-Thiokol report used data from only six launches (exactly a quarter of all launches up to that date) and for only two of those six was the launch temperature shown (and even then on a separate page).

In other words, the graphical process failed both at the analytical and the presentational level. At least in this sense it is difficult not to sympathise with the unfortunate Joe Kilminster, Vice President of Space Booster Programs at Morton-Thiokol, who rejected such narrowly based and incompetently presented advice and wrote to NASA, 'Temperature data not conclusive on predicting primary O-ring blow-by [...] MTI recommends STS-51L launch proceed on 28 January 1986' – 12 hours before Challenger exploded.

Curiously (and sinisterly, in view of the Columbia tragedy in 2003, in which similar communications failure also played a crucial role, as we shall see below) this graphical incompetence carried over into the report by the very Presidential Commission charged with investigating the disaster. Four months after the disaster the Commission report included a post-disaster Morton-Thiokol chart, summarising the previous history of O-ring damage, which made brilliant use of techniques drawn from the whole spectrum of graphical obfuscation. Had Morton-Thiokol been seeking to obscure the role that graphical incompetence had played in the disaster – let it be said immediately that there is no evidence that they had been – they could scarcely have chosen more wisely.

The modest CYA disclaimer at bottom left (known in the presentation trade as 'all's balls outside these walls') announces that the chart was prepared by professionals. The 48 rocket icons, for those of us who thought that 'boosters' was a misprint for 'roosters', document their professional opinion of their audience. But it's the unexplained symbols for types of damage (the legend, like the temperature information on the

History of O-Ring Damage in Field Joints (Cont)

O-Ring Temp (°F)											
66°	70°	69°	80°	68°	67°	72°	73°	70°	57°	63°	78°

| SRM No. | 1 A | 1 B | 2 A | 2 B | 3 A | 3 B | 4 A | 4 B | 5 A | 5 B | 6 A | 6 B | 7 A | 7 B | 8 A | 8 B | 9 A | 9 B | 10 A | 10 B | 11 A | 11 B | 12 A | 12 B |

O-Ring Temp (°F)											
70°	67°	53°	75°	67°	70°	81°	76°	79°	75°	76°	58°

| SRM No. | 13 A | 13 B | 14 A | 14 B | 15 A | 15 B | 16 A | 16 B | 17 A | 17 B | 18 A | 18 B | 19 A | 19 B | 20 A | 20 B | 21 A | 21 B | 22 A | 22 B | 23 A | 23 B | 24 A | 24 B |

INFORMATION ON THIS PAGE WAS PREPARED TO SUPPORT AN ORAL PRESENTATION AND CANNOT BE CONSIDERED COMPLETE WITHOUT THE ORAL DISCUSSION

Report of Presidential Commission on Space Shuttle Challenger Accident, 1986

evening before the disaster, is helpfully on a separate page) and, above all, the display sequence of previous launches that should inspire awed respect.

As we have seen, the fatal error the evening before launch lay in omitting three quarters of the data from both the analysis and the presentation report. Here, the data are complete, if almost illegible. But the telephone book school of category sequencing has arranged them by launch number, which is as useful for finding the data for a particular launch as it is useless for drawing any conclusions about the connection between temperature and damage.

History of O-Ring Damage in Field Joints

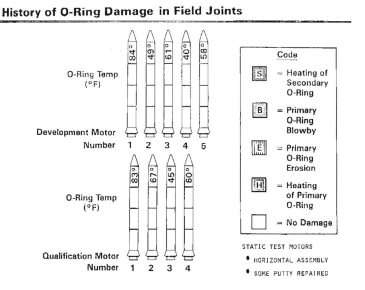

Report of Presidential Commission on Space Shuttle Challenger Accident, 1986

The Presidential Commission cruelly ripped the data from their iconostasis above and published a clumsy and offensively honest chart of O-ring damage as a function of temperature in the first volume of its report. The juxtaposition of a chart showing only flights in which damage occurred, still more complete than the data shown to management by the engineers, at the top of the page and a chart of all flights including those without damage at the bottom is unpleasantly explicit, sums up the Commission's key conclusion on a single page and seems almost calculated to put decent graphical deceivers out of a job.

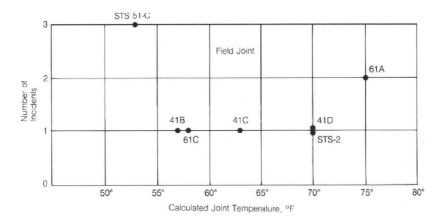

Figure 6
Plot of flights with incidents of O-ring thermal
distress as function of temperature

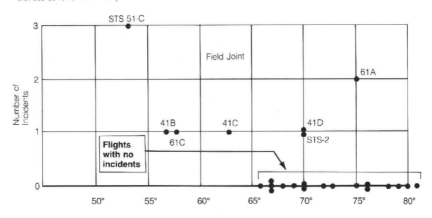

Figure 7
Plot of flights with and without incidents of O-ring
thermal distress

NOTE: Thermal distress defined as O-ring erosion, blow-by,
or excessive healing

Report of Presidential Commission on Space Shuttle Challenger Accident, 1986

In fact, as Tufte has pointed out, it could have been a lot worse and
the connection between temperature and damage made even more
striking. He redraws the data more explicitly (by using a more sensitive
and quantified damage axis) and by extending the scale to the left
(increasing its scope by over a third) to include the expected temper-

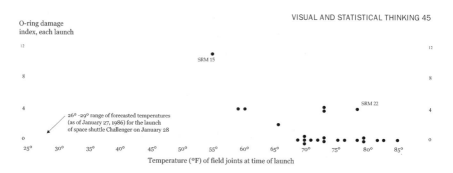

O-ring damage
index, each launch

SRM 15

26° –29° range of forecasted temperatures
(as of January 27, 1986) for the launch
of space shuttle Challenger on January 28

SRM 22

Temperature (°F) of field joints at time of launch

Tufte, *Visual Explanations,* 1997

ature of the Challenger launch. In this most poignant of might-have-beens, it is difficult to see how anyone who was in possession of the full facts, a neutral graphic and their own faculties could have recommended launching Challenger under prevailing weather conditions at the Cape. But, of course, no one at decision-taking level was.

Columbia: Challenger, the sequel

As Charlton Heston and the American National Rifle Association never tired of pointing out in defence of the liberalism, indeed absence, of US firearms legislation, 'Guns don't kill people, people do!' A similar defence can be made of Microsoft's astonishingly successful graphics software packages. The possession of PowerPoint and Excel, rather than a stencil and a rotring pen, no more guarantees graphic deceit than possession of a Kalashnikov, rather than a muzzle loader, leads ineluctably to mass murder. There is in logic a clear distinction between a necessary and a sufficient condition, which this book has no intention of blurring.

So the fact that the charts that failed to prevent the Challenger disaster in 1986 differed only from those that failed to rescue the Columbia astronauts in 2003 in being hand-written rather than composed in PowerPoint, is but a footnote to cultural history. The key point in both disasters was the quality of thought and graphic awareness that lay behind fatally ineffective communication.

Columbia was launched on 16 January 2003, at 10:39 local time: 17 years less 12 days after Challenger. After an otherwise normal mission, all seven astronauts on board were killed when Columbia broke up on re-entry over the southern United States on 1 February. Horrifying though this was, it was not entirely unexpected as parts of the spray-on foam insulation (SOFI) of the main fuel tank had been seen to break away soon after launch and strike Columbia's wing. Since soon after the launch NASA had been worrying that this collision had damaged the heat resistant ceramic tiles that protected the largely aluminium structure of Columbia from the fierce temperatures of re-entering the earth's atmosphere. They asked Columbia's builders, Boeing, to comment on the probability of the tile damage being severe enough to endanger the craft during re-entry.

This was no idle enquiry as there were two measures that NASA could have taken if the damage had been assessed as life threatening. First, a space walk by the crew of Columbia could have been organised not only to inspect the damage, which was not visible from anywhere within the orbiter itself, but also to carry out emergency repairs that would at least have reduced the probability of burn through and structural damage during re-entry. Such repairs would have been somewhat Heath-Robinson, involving for example placing heavy steel tools in any hole and cementing them in place with water frozen in the cold of space, but the period of maximum danger during re-entry was only of a few minutes' duration and even a large wrench might have deflected some of the heat.

Second, and unusually, another space shuttle, the Atlantis, was already in advanced preparation for a launch on 1 March. NASA estimated that by working round the clock, and if nothing went wrong during the count-down, Atlantis could be launched without omitting any of the normal safety checks by 10 February, giving a narrow window of five days for a rescue before Columbia was due to run out of breathable air on 15 February.

Boeing replied with a 13-page (again) document dated 23 January 2003 based largely on a previous case of tile damage to Columbia on mission STS-50 in 1992 and on the results of computer simulation of tile strikes called 'Crater'. The last 'Summary and Conclusion' page of the report is hedged around with conditions but ends '...safe return indicated even with significant tile damage'. Most attention in the subsequent investigation of the disaster focused however on page 6 of the Boeing report to NASA, which at first glance seemed to suggest that there were empirical grounds ('review of test data') for thinking that the damage to the tiles would not be life threatening. But as one descends into the tortured syntax of the body of the slide's text it gradually becomes apparent that it was the Crater simulation software that had been reviewed and had seemed to produce 'conservative' results (in other words, predicted greater damage than actually occurred), which is comforting until one reaches the last line on the chart. 'Volume of ramp is 1920cu in vs 3cu in for the test', when translated into English, means quite simply that Crater had only ever been validated on objects 1/640th the size of the chunk of foam suspected of having hit Columbia and that therefore Boeing did not have the slightest idea of what damage might have been caused to the tiles. On this basis the conclusion on page 13 should have read more simply 'Your guess is as good as ours' and should have provoked a burning desire at NASA to get one of the crew to inspect the hole (an alternative discussed but not followed up was to get military reconnaissance to photograph it) and patch it to the extent possible, and in the meantime to accelerate the launch preparation of Atlantis. But it takes a lot of courage to admit absolute ignorance to a major customer about a project involving national prestige.

Summary and Conclusion

- **Impact analysis ("Crater") indicates potential for large TPS damage**
  - Review of test data shows wide variation in impact response
  - RCC damage limited to coating based on soft SOFI
- **Thermal analysis of wing with missing tile is in work**
  - Single tile missing shows local structural damage is possible, but no burn through
  - Multiple tile missing analysis is on-going
- **M/OD criteria used to assess structural impacts of tile loss**
  - Allows significant temperature exceedance, even some burn through
    - Impact to vehicle turnaround possible, but maintains safe return capability

Conclusion

- **Contingent on multiple tile loss thermal analysis showing no violation of M/OD criteria, safe return indicated even with significant tile damage**

BOEING   2/21/03   13

NASA, 2006

Review of Test Data Indicates Conservatism for Tile Penetration

- **The existing SOFI on tile test data used to create Crater was reviewed along with STS-87 Southwest Research data**
  - **Crater overpredicted penetration of tile coating significantly**
    - **Initial penetration to described by normal velocity**
      - Varies with volume/mass of projectile (e.g., 200ft/sec for 3cu. In)
    - **Significant energy is required for the softer SOFI particle to penetrate the relatively hard tile coating**
      - Test results do show that it is possible at sufficient mass and velocity
    - **Conversely, once tile is penetrated SOFI can cause significant damage**
      - Minor variations in total energy (above penetration level) can cause significant tile damage
  - **Flight condition is significantly outside of test database**
    - **Volume of ramp is 1920cu in vs 3 cu in for test**

NASA, 2006

As in the Challenger disaster, the problem lay in communicating engineering knowledge to decision makers comprehensibly and powerfully. The PowerPoint format could have been forced to accommodate clear thinking without losing impact.

Dr Snow and the Soho pump handle

Dr John Snow's map of the geographical distribution of deaths from cholera in Soho in 1854 has become an icon of graphical historiography and epidemiology. But like that of Newton's apple, the story has grown in the telling and lost its most interesting features in the process.

For Snow, the cholera outbreak in Soho in late August 1854 caused by contamination of water from the pump at the corner of Broad Street and Cambridge Street was at first a diversion from his investigation of the *piped* water supplied by various water companies in London, itself part of his more general research into the 'Mode of Communication of Cholera', as his monograph in autumn 1854 was called. The debate was only finally concluded by Robert Koch's identification of the bacterium *Vibrio cholerae* in Egypt in 1883. In 1854 the battle was between those who blamed airborne 'vapours', possibly released from ancient plague pits by the excavations necessary to install modern sewers, and those such as Snow, who had become convinced that the cholera agent, whatever it was, was strictly waterborne. In two papers in 1849 Snow had adduced clinical evidence to suggest that since cholera attacked the intestines rather than the lungs the waterborne theory was more likely. In 1854, before the Soho outbreak, he was comparing the medical fates of different water companies' customers but his conclusions, though later shown to be correct in principle, were unconvincing at the time. He failed properly to take account of the size of each company's clientele and of multiple sources of water. The famous Soho map and the infamous Broad Street pump are but a small part of his efforts to demonstrate waterborne cholera transmission.

His main scientific opponent was William Farr, who considered that water was just one of several different transmission pathways including air and water. So the difference of opinion between Snow and his

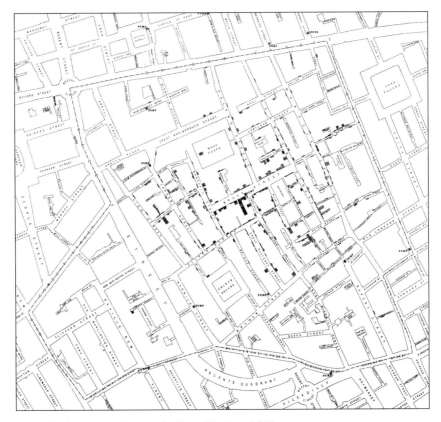

Snow, *On the Mode of Communication of Cholera*, 1855

contemporaries was less dramatic than is often suggested and his personal relationship with Farr remained excellent and intellectually fruitful, with constant exchanges of new data and theories until Snow's death in 1869.

Both Snow and Farr relied heavily on statistics – Farr, originally a doctor, had become Statistical Superintendent of the General Register Office and improved the collection of health and mortality statistics, though only Snow seems to have turned his statistics into charts. Yet both men are best remembered for statistics that we can now see supported their opponents' theories at least as well as their own.

Farr developed a formula, derived from his airborne transmission theories, connecting height above sea-level inversely with the incidence of cholera. Applied to 1850s London, his formula produced astonishingly accurate predictions of relative cholera incidence in different areas of the city and was confirmed by further investigations in Liverpool. A chart comparing actual cholera mortality and the predictions of Farr's Law is shown here (Farr himself used a simple table) and seems to offer startling support for his theory. It also coincided neatly with the observation that cholera tended to be most common in seaports. Of course, we now realise that seaports were particularly at risk simply because

Comparison of actual and predicted cholera mortality per 10,000 population in London by height above high-water mark, 1849

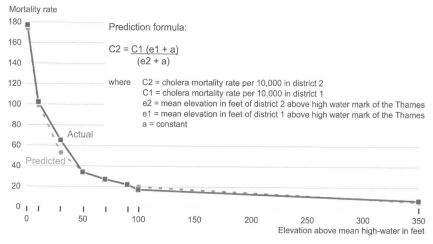

Prediction formula:

$$C2 = \frac{C1 \, (e1 + a)}{(e2 + a)}$$

where
C2 = cholera mortality rate per 10,000 in district 2
C1 = cholera mortality rate per 10,000 in district 1
e2 = mean elevation in feet of district 2 above high water mark of the Thames
e1 = mean elevation in feet of district 1 above high water mark of the Thames
a = constant

Data from: Eyler, *The changing assessment of John Snow's and William Farr's cholera studies*, 2001

cholera is waterborne; water together with its impurities flows downhill and you can't get further downstream than a seaport. It is also generally true that the higher the elevation, the lower the risk of cross contamination of water supplies and finally that, at least in 19th century London, the upper classes, who for a host of reasons suffered less disease, tended to live higher up the sides of the Thames valley than the lower classes. But nothing in Farr's comparison contradicts either the airborne or the waterborne theory.

Snow used two parallel chains of argument. The best known, and unfortunately then as now the most striking, was his map of cholera deaths shown above. But on its own, the map proves nothing at all except that the deaths were geographically concentrated. This certainly suggests a shared cause of determinate position but tells us nothing about whether the cause was air or water borne. As Snow's contemporary critics rightly pointed out, any such group of deaths in 1850s London would have been found to cluster around a water pump. There were simply a lot of water pumps about. The deaths also clustered around the site of old plague pits, a suggested source of airborne infection, and of which London also possessed large numbers.

It's possible that Snow used a preliminary, and of course highly incomplete, version of the map when he went to St. James's Parish Council on the evening of 7 September, 1854 and persuaded the vestry-

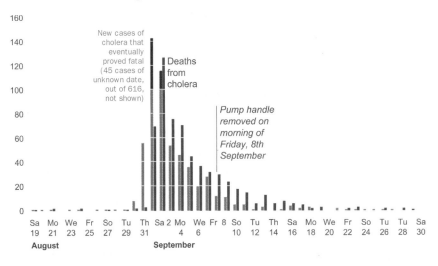

Data from: Snow, *On the Mode of Communication of Cholera,* 1855

men to have the handle of the Broad Street pump removed. This was done early the following morning and the epidemic petered out within ten days. But it is highly unlikely that removal of the pump handle had the slightest effect on the epidemic. There were only ten new infections on Friday, 8 September (given the short incubation period, the last day before the effect of the removal of the pump handle could have become noticeable), down from 144 at the peak in the previous week. As we now know, this was because most of those who lived near the Broad Street pump were either already dead (the epidemic claimed some 616 lives in only a few city blocks, housing on Snow's calculations 3,000–4,000) or had fled. The remaining population was either immune to cholera (a marked tendency, for example, in those with blood of type AB); had developed temporary immunity from recent mild exposure, or habitually used water from other sources, such as that from a deep well at the brewery or piped water at the workhouse.

It is on Snow's investigation and description of the last two groups (anomalous negatives) and of individual cases of cholera among those who had fled or even, in one famous instance, left the area months before (anomalous positives) that his reputation as one of the earliest and greatest epidemiologists should rest. Snow found no severe cases of cholera had appeared among more than 70 workers at the brewery and only five among 535 inmates at the workhouse, both of which had their own water supplies. Had cholera been communicated other than by water, say by *miasmata* in the air, over 120 deaths should have occurred among workers and inmates. By following up and investigating individual cholera deaths as far away as Hampstead, way beyond the then accepted operational radius of *miasmata*, Snow was able to show in all cases that the deceased had recently drunk water from the Broad Street pump, prized by some former inhabitants for its distinctive flavour.

Snow's fabled map is more interesting to us as a powerful example of how graphical sophistication, rightly used, can distract from and even fatally undermine a completely correct case.

In terms of public health, the partial disagreement between Snow and Farr probably had limited impact. In the years after Snow's death, Farr gradually adopted the exclusively waterborne hypothesis and in practical terms the measures required to eliminate the entirely mythical airborne miasmata arising from rotting vegetation, rubbish and faeces overlapped considerably with those required to eliminate contamination of water supplies, namely modern sewerage, rubbish collection and improved housing.

Conclusions for those who practise to deceive

First the obvious: getting graphics right, or depending on your point of view, wrong, matters. Graphics can kill.

Secondly, and slightly less obviously, graphical power belongs to him who gets in firstest with the mostest. The power of a graphic to distort or obscure the truth often depends on how early in the development of the chart we can get our hands on the data, long before we put graphic ink to paper. This is because one of the most effective deception agents is removal of data perspective. In all three cases, Challenger, Columbia and Cholera, data was taken out of perspective, at least partly for graphical reasons.

19 of the 25 shuttle flights before and including Challenger were ignored, fatally reducing the impact of available temperature data. Even the subsequent enquiry destroyed perspective by getting data sequence wrong.

The size of the actual foam insulation debris relative to the hopelessly restricted Crater simulation data base, and hence ignorance of the size of the problem, was graphically obscured, condemning Columbia to burn up on re-entry.

Snow's map showed only cholera deaths near the pump. The crucial point was cases of survival near the pump and cases of death far from it. Luckily his incompetent graphic was largely discounted at the time.

Having established that graphics matter and seen a few examples of how to deceive and obscure with their help, it is now time to get to know in detail the weapons of graphical manipulation at our disposal for what I like to think of as strictly pre-emptive distortion. If we don't get in there and bend the data, someone else will.

3 DISTORTING VALUES

Manipulating the Data

I f you can get to the data in time, deceit is easier, and also harder to detect, if you manipulate the numbers themselves rather than the way they are displayed graphically. Of course, in a sense, the number of possible data deception techniques is limitless, but our self-imposed restriction of avoiding outright lies and sheer practicalities, given that the numbers to be manipulated already exist, narrows it down a lot. In practice, there are two groups of common data deception techniques at our disposal before getting into the graphics: condensation fraud and unit fraud.

Condensation fraud

This is not a statistics textbook, but it's useful to remind ourselves of the charmingly exploitable weaknesses of that key statistical condensation tool, the average. All averages are ways of condensing our knowledge about groups of differing individuals or objects. For homogeneous populations, to describe one member is to describe all; no average is required. So in a way, all averages are, by definition, less than the full truth. Averages can be used for deception either by hiding unwanted data within them, by failing to disclose what sort of average they are, or by careful selection of the members of the averaged group.

Using averages to hide important variations

If you want to hide data, try putting it into a larger group and then use the average of the group for the chart. The basis of the deceit is the endearingly innocent assumption on the part of your readers that you have been scrupulous in using a representative average: one from which individual values do not deviate all that much. In scientific or statistical circles, where audiences tend to take less on trust, the 'quality' of the average (in terms of the scatter of the underlying individual figures) is described by the standard deviation, σ, although this figure is itself an average.

Take an equally standard British obsession: is the euro good or bad for economic growth? Few people on either side of the euro debate are still open to arguments based on mere facts, but here they are, for the sake of the example.

There doesn't seem to be much to choose between euro and non-euro states in terms of real GDP growth since the introduction of the euro. The five fastest growing states use the euro, as do the four slowest growing. The middle ranks, sixth to eleventh, contain three

EU-15 states using €/not using €)	Real GDP growth % per year					Rank %
	2002	2003	2004	2005	2001–2005	
Ireland	6.1	3.7	5.4	4.9	21.6	1
Greece	3.8	4.7	4.2	2.9	16.5	2
Luxemburg	2.5	2.9	4.5	3.8	14.4	3
Finland	2.2	2.4	3.7	3.3	12.1	4
Spain	2.7	2.9	3.1	2.7	11.9	5
UK	2.0	2.5	3.2	2.8	10.9	6
Sweden	2.0	1.5	3.6	3.0	10.5	7
Belgium	0.9	1.3	2.9	2.2	7.5	8
Austria	1.2	0.8	2.2	2.1	6.4	9
France	1.2	0.8	2.3	2.0	6.4	10
Denmark	0.5	0.7	2.4	2.3	6.0	11
Italy	0.4	0.3	1.2	1.2	3.1	12
Germany	0.2	0.0	1.6	0.8	2.6	13
Netherlands	0.6	−0.9	1.4	1.0	2.1	14
Portugal	0.4	−1.1	1.0	1.1	1.4	15

Eurostat, 2005

euro and three non-euro states. It looks as though something other than euro membership or non-membership is mainly responsible for the differences in rates of economic growth in Europe since 2001.

But this is defeatist. The solution lies in the choice of suitable averages, in this case arithmetic means that will hide the inconvenient data and encourage us to ride, unburdened by factual detail, to the rescue of the pound. The list of states above, in order of real growth, gives a clue on how to fight back. Two of the low growth euro economies are very big, Italy and Germany, and two of the high growth economies, Ireland and Luxemburg, very small.

Sure enough, the averages of the euro and non-euro groups give a much clearer picture. Indeed, there seems little room for discussion. EU states within Euroland grew on average only half as fast as those outside from 2001 to 2005. Game, set and match to those plucky little local currencies!

	Annual real GDP growth (%)				GDP growth 2001–2005
	2002	2003	2004	2005	
Eurozone average	0.9	0.7	2.0	1.6	5.3
Non-€ average	1.9	2.2	3.2	2.8	10.4

Eurostat, 2005

The *Independent on Sunday*, like the *American Business Week*, went one better and compared the euro zone average with individual values for Sweden and Britain. Curiously, the third non-euro member state, Denmark, with roughly the same growth rates as Euroland, failed to make the cut in either publication. Not a lie has been told in arriving at either of the two radically different and strictly quantitative views of the merits of the euro presented here. With the minor exception of omitting the Danish figures, the *Independent* cannot even be accused of biased selection of data.

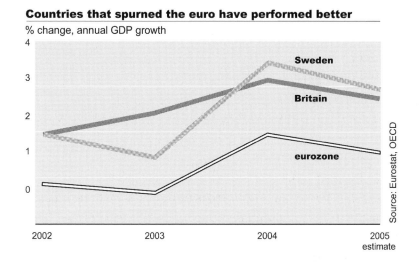

Countries that spurned the euro have performed better

% change, annual GDP growth

Source: Eurostat, OECD

Independent on Sunday, 2005

It's difficult to put a PDQ (potential deceit quotient – see Introduction) on this example, but considering that a 'doesn't make much difference either way' has been turned into a cast-iron 2:1 advantage, the PDQ must itself be around 2:1. The STD (sore thumb discount) is as near to zero as makes no difference, giving a net expected value also of 2:1 – a technique to treasure, not only in the Eurodebate.

Defining averages unusually

There's surprising mileage in camouflaging the definition of 'average'. The most notorious example is 'Most people have an above-average number of legs' (which is only true in the sense of 'mean' rather than 'mode' or 'median'; think about it), but subtler variations can put whole nations in a bad light.

SOME ARE MORE AVERAGE THAN OTHERS*

Mean = $\dfrac{\text{total of all observed values}}{\text{number of observations}}$

Mean *number of legs per person = 1.998*

Median = *observed value of which half the observed values are greater and half smaller*

Median *number of legs per person = 2.0*

Skew = *median – mean = 0.002*

Mode = *most frequently occurring observed value*

Mode *number of legs per person = 2.0*

Mid-range value = *observed value equidistant from minimum and maximum observed values*

Mid-range *number of legs per person = 1.0*

**In a standard, Gaussian, distribution all four values coincide.*

GDP per head is often used as a measure of a country's economic success compared to others. It is a mean, the result of dividing GDP by the number of people in the country. Its quality as a description of national wealth depends just as crucially on the distribution of incomes within the country as the usefulness of the average growth rates above depends on the variation within the euro and non-euro groups.

The greater the disparities of wealth, the less reliably GDP per head describes reality. We would hesitate to attribute success to an economy in which one person was immensely wealthy and all others lived in abject poverty, even if the average (mean) GDP per head were very high. This problem is getting less and less theoretical because income inequalities have been growing throughout the Western world, particularly in the USA and UK, for 20 years. If the disparities get much bigger, we shall probably have to start using median incomes rather than mean GDP as the basis of comparison for economic success.

Unit fraud

All quantitative data has to be expressed in some sort of unit. The choice of which unit is often the place to start if we are seeking to deceive.

Decisions about which units to use include whether to use absolute, ratio, differential or index values, whether to weight ratio values to reflect the relative importance of the category (for example comparing

British and Luxemburg birth rates) and whether to use annual totals or cumulative values. There are other handy deception opportunities in sheer semantic confusion about the precise definition of some units, 'productivity' for example.

All these choices involve seriously attractive PDQs and generally favourable STDs. The STDs are low because there cannot be universally valid rules about when to use, say, a ratio as opposed to an absolute value or time period rather than cumulative totals. The right choice is entirely and enticingly dependent on context, above all on the message the chart is designed to communicate. In other words, it is nearly always possible to cobble together some sort of argument to support your choice of value unit and anyone contesting that choice tends to look like a medieval cleric disputing the number of angels that can dance on the point of a pin. The choice is only really limited by the imaginative power of the author.

Using absolute instead of per capita values

In the last quarter of a century the British Crime Survey has recorded a steep rise followed by a steep fall in the violent crime rate in England and Wales. In 2003/04 rates were 640 per 10,000 adult inhabitants, an increase of 15% over 558 in 1981 – a significant increase but scarcely the approaching end of civilization as we know it, especially in view of the decline of almost half in the last ten years from 1,046 in 1995. After similar peaks in the mid-1990s most other sorts of crime are now roughly back at 1980 levels. Even violent crime, after a nasty peak in 1995, is now back roughly where it was in 1980. This does not make for very gripping headlines, even less for the mobilisation of opinion against the government in power since 1997.

However the best of a weakish case can be made by changing the units from crime rate to crime incidence (from figures per 10,000 population to absolute numbers of crimes) to take advantage of the population increase, roughly 7%, since 1981. Using a bit of graphic distortion as well, placing the violent crime figure since 1981 immediately below a similar chart showing the general decline in all crime since 1995, makes the violent crime rise since 1981 look that much steeper. No one can object, as both horizontal axes are clearly labelled and any

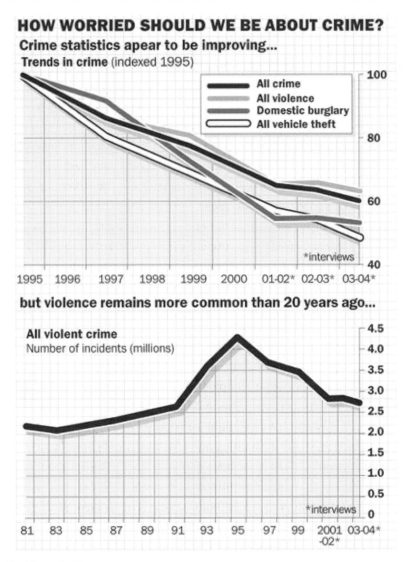

HOW WORRIED SHOULD WE BE ABOUT CRIME?

Crime statistics apear to be improving...

Trends in crime (indexed 1995)

- All crime
- All violence
- Domestic burglary
- All vehicle theft

*interviews

100

80

60

40

1995 1996 1997 1998 1999 2000 01-02* 02-03* 03-04*

but violence remains more common than 20 years ago...

All violent crime
Number of incidents (millions)

*interviews

4.5
4.0
3.5
3.0
2.5
2.0
1.5
1.0
0.5
0

81 83 85 87 89 91 93 95 97 99 2001 03-04*
-02*

The Times, 2005

influence of the top chart on the one below is presumably accidental. The result is not bad at all, given the unpromising nature of the original data, though it could be improved further by starting the comparison in 1983 and stopping in 2002/3, giving a rise in the number of violent crimes of over a third, instead of the modest 15% rise in rates we started with. The combined effects of data selection and unit fraud give a PDQ of around 1.3:1 in *The Times* version.

PDQs in similar cases obviously depend both on data variability, to give scope for selecting a favourable time period, and on the length of the period, to give time for the difference between absolute and per capita values to grow. The latter has only been good for about 7% every 20 years since 1945 and so the technique is of only modest power on its own. The related technique of using nominal rather than purchasing power parity currency values (to exaggerate growth) is four or five times as powerful, as inflation rates have tended far to outstrip population growth.

As a very general rule, absolute values are less likely to lend themselves to deceit than ratio values, such as per capita figures. Apart from anything else, ratios, by definition, have twice the number of variables to manipulate as absolute values. More importantly, absolute values have the most direct relationship with reality and so tend to keep statistical feet on the ground. But this only hinders us in contexts where readers and viewers possess background knowledge which will help them to put our values into perspective. If this background knowledge is lacking, even, perhaps especially, absolute figures may be used to change perceptions of the world.

Take for example the AIDS epidemic in sub-Saharan Africa and its effect on children under 15. This map published by the *International Herald Tribune* seems at first sight to be repeating an all too grimly familiar tale. Forget the distracting and absurd row of figures across the top for the benefit of those who have difficulty understanding the phrase 'less than one in 20' and concentrate on the map itself.

The eye is caught, no doubt intentionally, by the darkest shading, for Nigeria, the Democratic Republic of the Congo, South Africa, Zimbabwe, Tanzania and Ethiopia. The message is clearly that these are the countries with the most heart-breaking problems of HIV positive children.

In a sense, this is not quite untrue. These are indeed the countries with the largest known absolute numbers of infected children in Africa. But a glance at the one exception, Lesotho, with relatively light shading, gives the game away. Lesotho has the greatest problem relative to population under 15 of any of the countries for which numbers are quoted. With a prevalence rate of 3% its children are twice as badly stricken as South Africa's with 1.5% and six times as badly as the Democratic Republic of the Congo's with 0.5%, whose shading dominates the chart.

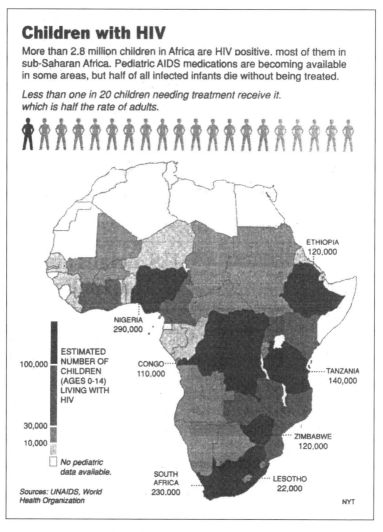

Children with HIV

More than 2.8 million children in Africa are HIV positive. most of them in sub-Saharan Africa. Pediatric AIDS medications are becoming available in some areas, but half of all infected infants die without being treated.

Less than one in 20 children needing treatment receive it. which is half the rate of adults.

ETHIOPIA
120,000

NIGERIA
290,000

100,000 — ESTIMATED NUMBER OF CHILDREN (AGES 0-14) LIVING WITH HIV

CONGO
110,000

TANZANIA
140,000

30,000

10,000

ZIMBABWE
120,000

No pediatric data available.

Sources: UNAIDS, World Health Organization

SOUTH AFRICA
230.000

LESOTHO
22,000

NYT

International Herald Tribune, 2006

What seems to have happened is that figures have been hijacked from one purpose to another. Figures that might just have had a certain specialist relevance for aid officials trying to decide how much of a limited budget for children's health for the whole of Africa to devote to any one country, have been co-opted to serve as indicators for the seriousness of the epidemic in each country. To put it another way, this map, by taking absolute numbers rather than per capita rates of HIV infections, makes our concern for African children entirely dependent on the size of the populations of the countries in which they happen, just, to live. It's difficult to think of a less appropriate way of focusing public concern than this complete reliance on the way African frontiers were fixed at various European congresses a century ago. It brings to mind the expostulations of a senior American envoy, when taxed on the BBC's Today programme with the fact that the USA provided less than half as much development aid as the EU, that if you added up the individual US states you'd get a pretty big figure too.

This seems to be just one example of a general preference for absolute figures among no-nonsense, both-feet-on-the-ground Yankees. It still applies when the figures require considerable additional knowledge and quite some mental arithmetic on the part of the reader to tell their full story. The preference is somehow lovable and points the way to some interesting distortion possibilities.

Take as simple an example as the geographic distribution of ex-servicemen and women ('veterans', as the American language more succinctly puts it) living in the USA. A moment's thought suggests that anyone who is ex anything is probably older than people who aren't and

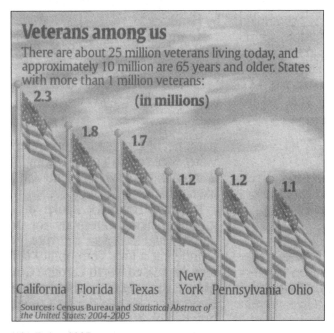

Veterans among us

There are about 25 million veterans living today, and approximately 10 million are 65 years and older. States with more than 1 million veterans:

(in millions)

California	Florida	Texas	New York	Pennsylvania	Ohio
2.3	1.8	1.7	1.2	1.2	1.1

Sources: Census Bureau and *Statistical Abstract of the United States; 2004-2005*

USA Today, 2005

should therefore tend to congregate in the sunny retirement states of the American southeast and southwest. As it turns out, nothing could be further from the truth.

In terms of the proportion of veterans in the total population, Nevada and Florida only come in at ranks 5 and 6 and Arizona at 23, the top three spots being taken by states connected only by their lack of oranges and deckchairs: Maine, Montana and West Virginia. At the other end of the scale, New York, California and Utah rub shoulders in the last three places. Pennsylvania and Ohio rank at 18 and 20 respectively. So veterans' retirement choices are anything but sunny side up. Upon reflection, the distribution probably has more to do with the places veterans would have tended to serve last and hence with the size of local military bases. Military pensions are not of a value to support very free choice.

But this sort of speculation is stifled at birth by the way that *USA Today* chose to present the figures for states with more than one million veterans. By using absolute figures, information has effectively been turned into factettes, of little value unless they are reconnected with other numbers, most obviously total population of the states concerned.

The lesson for those who practise to deceive is that making the right choice between absolute and ratio figures can reduce information content almost to zero without embarrassing reductions in the number of figures on the chart. Throw in, as here, a surreptitious stretching of the flagpoles to understate differences – California's Old Glory should be floating more than twice as high as Ohio's – and the ink to information ratio falls not far short of the infinite.

Using unweighted rather than weighted values

We tend automatically to think of all the categories represented on the horizontal axis of a column chart as being equally important. They vary of course on the value axis, otherwise there would be little point in the chart, but there is somehow this feeling that they are in other respects similar members of a group. This convention can be put to good use to manipulate the message of the most boring bar or column chart.

Do you think that the euro zone is doomed, not least because its members' growth and inflation rates will diverge too far from each other,

Gap between eurozone and individual member country growth rates in %

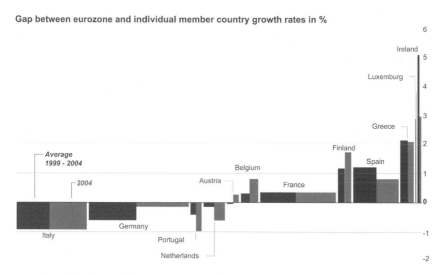

Data from: *The Times,* 2005

and from the average of the zone, for Euroland to hold together? If you do, you will probably find this chart for growth rates in the euro zone disappointing. The height of the columns in the chart shows each economy's divergence from the euro zone average; column width is proportional to the size of the respective national economy in 2004. The vast majority of the shading is within 1% of the zone average and areas more than 2% from the average are so small as to be difficult to draw. In other words, divergence seems to have been modest; so modest as to be less than that among counties, states or regions within other, older established currency areas like the dollar or sterling.

The cure for this disheartening message is very simple. Most of the states diverging dramatically from the others are small or very small. If column width is adjusted to reflect sovereignty, which of course is 100% in all cases, Greece, Luxemburg and Ireland are restored to their rightful place in the EU pantheon and the differentials loom much larger, even life-threateningly. Note that at least two-thirds of the red and black area in the upper, growth rate chart, is contributed by the three smallest economies in the euro zone.

The PDQ involved in assigning equal visual weight to UK and Luxemburg ratio figures is about 120:1. For Germany:Luxemburg the PDQ is roughly 165:1. There is almost no STD to worry about. This is

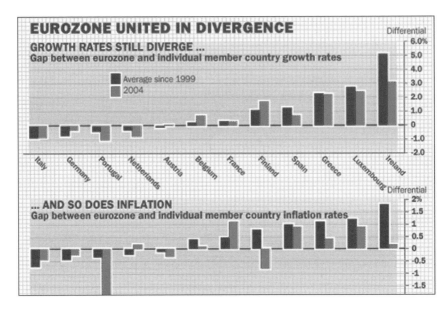

The Times, 2005

partly because we are so used to the bar and column equality convention that it scarcely registers. Equally importantly, the less misleading, area-proportional, alternative has to be hand drawn and so remains rare. It is not offered by any of the mainstream graphics packages. But it's quite a technique if the underlying figures are favourable!

It's difficult to say whether non-weighting fraud is really an example of manipulating the data before it gets as far as the chart or a graphical manipulation that takes place on the chart itself. It has been included among these pre-graphical techniques because its use involves a conscious decision not to let half the relevant data get as far as the chart.

Using cumulated totals

Deceit through cumulation is a bit of a golden oldie, but still frequently used in some highly respectable publications.

Take for example the worldwide HIV epidemic, an entirely justified focus of worldwide attention, sympathy and alarm, not least because it seems to afflict most severely those societies least able to defend themselves. The figures are well known, but bear repeating here. It is of course immediately apparent that, appalling though these figures are – we are talking about a

Annual new HIV infections and AIDS deaths worldwide 1980 - 2004 (millions)

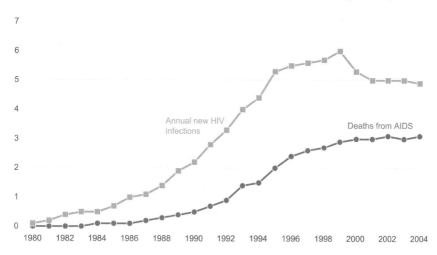

Data from: Worldwatch Institute, *Vital Signs,* 2005

Holocaust-sized catastrophe every two years – they do offer grounds for very cautious optimism. Since the turn of the millennium both annual mortality from AIDS and new HIV infections seem to have flattened out. New HIV infections have even decreased significantly since a peak in 1999.

It seems extraordinary that anyone should feel the need to supply added zest to such grim figures, but where there's a will there's a way and the simplest way is cumulation. The charts are from *Vital Signs 2005*, damned with faint praise by the head of the UN Environment Programme as 'a book of clear analysis and possible solutions'. They are a striking tribute to how cumulative figures can dramatise. The shape of the curves just screams 'this is a runaway problem'. The fact that the curves look almost parallel and deaths only lag infections by a couple of years also does its bit to encourage panic at the expense of cooler ratiocination.

The trick is twofold. It mainly depends on the fact that most curve diagrams depict annual figures. When cumulated figures are depicted there is often some sort of explicit visual warning, for example in the form of displaced stacks of columns like tottering piles of crates. In this sense the cumulated curve is probably more strongly self-limiting than most other manipulation techniques. The more it is used, the less powerful it becomes.

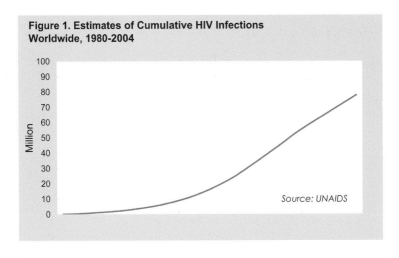

Figure 1. Estimates of Cumulative HIV Infections Worldwide, 1980-2004

Source: UNAIDS

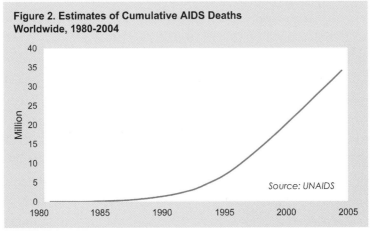

Figure 2. Estimates of Cumulative AIDS Deaths Worldwide, 1980-2004

Source: UNAIDS

Worldwatch Institute, *Vital Signs*, 2005

Cumulation offers another minor but sometimes worthwhile distortion in that cumulative curves tend to suppress short-term variations, indeed this bluntness is often the stated reason for their use. Looking very carefully indeed at the infection and death curves above, there is a barely perceptible flattening after 2000 but this is lost in the sheer size of the cumulated figures.

Of course, there are occasions when a cumulative figure is more useful than an absolute. The cumulated total of HIV positive survivors, for example, is crucial for health care planning, but neither of the charts above gives this figure and the chapter in which they are used is more concerned, quite rightly, with prevention, for which cumulation offers a dramatic, if shaky, foundation.

The PDQ is difficult to calculate, but the cumulative curves carry a subliminal suggestion that the problem is still growing as fast as it was in the mid-1990s, which would have made it about twice as big as it is now. PDQ is thus around 2:1. The technique is so old that a discount factor of at least 30% applies, giving an STD of around 0.7 and so a net value of about 1.4:1. This is useful, if not great.

Using change or delta values instead of absolutes

If inflation decreases, do prices go up or down? Is the axis label in the first example chart in this chapter '% change, annual GDP growth' strictly accurate? Surely it shows change in GDP in %, not changes in GDP *growth* in %, which would be the difference between one year's growth and that of the next year? If this seems nitpicking, so much the better; it will reduce the discounts applicable to the use of this handy deception technique, which relies on using change, or delta, values rather than absolutes.

Public and media attitudes to consumer price inflation differ from those to asset price inflation. Consumer price inflation (aka inflation): bad; asset price inflation (aka house prices and the FTSE or Dow Jones):

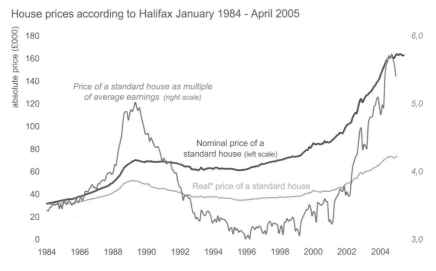

House prices according to Halifax January 1984 - April 2005

Data from: Halifax plc, 2005

UK House Prices
Year change %

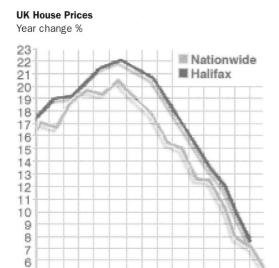

SOURCE : Halifax & Nationwide

BBC 2005

good. According to the Halifax, house prices in Britain in 2005 were on average just over five times their 1984 level in nominal terms, almost two and a half times in real terms after taking account of consumer price inflation and only about one and a half times compared to average wages.

On the basis of these figures for the last 20 years, you can see why the building societies are still relatively relaxed about the risks and consequences of future price collapse. In terms of price/earnings ratios and thus of default risk, prices – and hence mortgages – only moved above late 1980 levels a couple of years ago. As disposable incomes have risen even faster than average wages, the risk is still very moderate.

So what can we do to generate a healthy and media-worthy fear of house price catastrophe?

The BBC found an answer in simply taking year on year monthly change rather than absolute prices as above. The results look uncannily like the ballistic path of a market on its way to final impact in the Slough of Despond.

A nice touch is the broken, but not marked as such, value axis, which not only sharpens the angle of descent but also visually suggests final splat down sometime in the next six weeks. The BBC chart feels like a decline of a good four-fifths, while the absolute figures above, that barely perceptible wiggle in the last three or four months of the nominal and real price curves, signal no very significant change. So we're looking at a PDQ of 3 or 4 to 1 with a very moderate STD: nice one.

Using index values instead of absolutes

Index values also primarily show change, but with reference to one particular point in the past rather than the moving base line of change on a year or month ago. They can be manipulated in much the same way as year-on-year change values, but they do have an added ingredient, based on the fact that they look very like ordinary x/y line charts showing absolute values. So they create the impression that the value that has grown fastest is also the biggest, which of course is sometimes, but not always, the case.

Consider the figures on how the Government Actuary thinks the age structure of the UK population is going to change over the next 30 years, thought by many to hail the biggest threat to our standard of living since the Black Death. The proportion of people of working age will fall from today's 66% to only 59% in 2034, while the proportion of pensioners will climb from 16 to 25%. Putting it another way, each dependant (under 16 or over 65) will only be supported by less than 1.5 active workers as opposed to almost 2 today. This is quite a change and the simplest change diagram shows it clearly, if boringly.

UK population by age group, 2004 & 2034

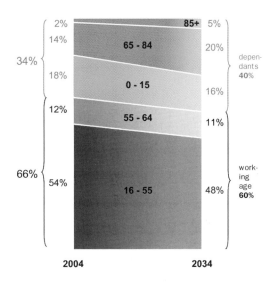

Data from: Government Actuary's Department, 2005

But this information on age structure is supposed to be the clarion call to awaken us to the end of civilization as we knew it. Nothing is supposed ever to be the same again. This is not quite the message of the sober chart here.

If we want to add bite to the Government Actuary's numbers, using indexed figures is one of the easiest ways. Instead of showing the before and after proportions, we can use index figures to show the change in each age group. As we are talking of a period of 30 years, the change in each index can be quite startling. In this example from *The Times* it looks as though the seriously wrinkly, the over 85s, are

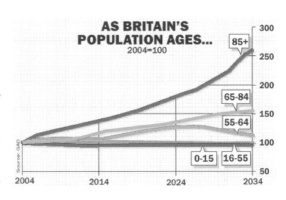

The Times, 2005

taking over the country, with a little help from the 65–84-year-olds. The chart artfully hides the fact that, although the number of 85+ year-olds goes up over two and a half times, the 2004 base is a mere 1.1m, or roughly 2% of the population. This is so low that even in 2034 only about 5% of the population will be over 85. The bulk of the population, the 16–55-year-olds still accounting for 48% of the population in 2034, is left lurking in the bottom sixth of the chart, partly thanks to a neat, unsignalled, broken axis.

Using unusual value definitions

Before leaving unit fraud, it is worth looking at the possibilities that are opened up by unclear definitions of units, especially when they tend to contradict everyday usage. The ideal situation, from which the example here is drawn, is when one word is used in different ways by the two sides of an argument.

HM Treasury has an enviable reputation as a test lab for the not quite untrue government statement, where facts are subjected to stresses almost unimaginable to the layman to discover what it takes to sever the last link to reality. The results of these experiments are published annually as the 'Pre-budget Report'.

As the graphic makes clear 'productivity' might mean 'output per hour worked', 'output per worker' or even 'output per person of working age', though the last stretches normal, even official, usage to the self-destruct design limit. The interesting point is the power of definition to distort geopolitical realities in the whole of the North Atlantic.

Chart b: International comparisons of efficiency, 2003

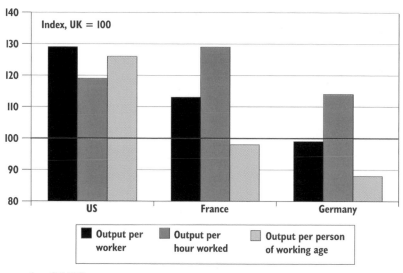

Index, UK = 100

Source:ONS, HM Treasury
Note: Data for 2003 are provisional; output per hour worked are experimental data.

Her Majesty's Treasury, *Pre-Budget Report*, 2004

In output per hour worked, the USA comes a poor second to France and only slightly ahead of Germany, which still suffers from the after effects of merging with 17 million East Germans in 1991. UK productivity is 12–25% lower. Only in terms of output per person of working age is UK performance acceptable, which may go some way to explaining why this eccentric definition of economic productivity has been included here. The problem with this unconventional definition is that it depends more on the percentage of working age people in work than on their productivity once they get to work. It doesn't even depend as much on unemployment as it does on all the other reasons why people may not be part of the workforce at all: being at school or university, looking after children or aged relatives and so on.

This is an outstanding example of supporting dodgy unit definitions with other devices such as indexed values, broken axes and poor, over-inked layout in a sort of amphibious, joint services assault on reality to produce an estimable, well-rounded chart. Comparing the lengths of the different coloured columns (in other words the differences in relative position stemming from different definitions) reveals PDQs up to 4.0:1.

The applicable STD depends on your view of the veracity of HM Treasury, on which this book cannot possibly comment.

Using nominal instead of real money values

Ever since the infamous inflation of the late 1920s, which played into the hands of a scruffy Austrian immigrant painter called Adolf Hitler, few countries have been as conscious of the perils of debauching the currency as Germany. To this day memories of this catastrophe are said to remain so strong that Germany insisted on the Growth and Stability Pact at the European Council meeting in Amsterdam in June 1997. The pact was supposed to stop garlic-flavoured governments running up colossal budget deficits – and hence government debt – at the expense of more fiscally sober Northern European states. It set a limit of 3% of GDP on new government borrowing in any one year, and was as plausible as a pint of Perrier in a Prohibition speakeasy, not least because many of its signatories had broken this limit in the preceding five years and many were to do so again before the ink was dry. But it was the price that the German public was thought to demand for abandoning the Deutschmark. Today most of the EU's bootleg government deficits are German and likely to remain so for several years. Though in fairness it should be pointed out that German deficits are harmless compared to the world's leading fiscal inebriates, Japan and the USA.

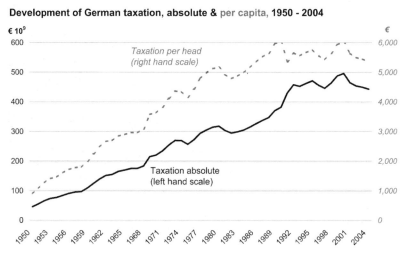

Development of German taxation, absolute & per capita, 1950 - 2004

Bundesminister der Finanzen, *Steuereinnahmen nach Steuerarten*, 2005

Against this background you might expect Germans to be acutely aware of the difference, the gulf, between nominal and real financial values over periods of more than a few years. But examples from the German media, presumably in tune with their readers, suggest that this may be optimistic. Take for example the way the tax burden was recently depicted in Germany's leading weekly, *Der Spiegel*.

Based on government figures the total tax take in Germany rose fairly steadily from soon after the founding of the Federal Republic in 1949 to the early 1990s and has stagnated since. The same is broadly true of taxation per head. All figures are expressed in real values, in this case 2004 euros. The picture in most other rich countries over the same period is similar and so hardly likely to quicken the journalistic pulse of a fearlessly investigative organ like *Der Spiegel*.

There are several techniques that could have been employed to make the figures more frightening. For example the annual figures could have been expressed in a different currency each year. With a little ingenuity and a list of 55 currencies arranged in descending order of the current exchange value of one currency unit, it would have been possible to increase the slope of the taxation curve more or less at will. Expressing the 1950 value in sterling, 1968 in US-dollars and 2004 in yen, and filling in the years in between with other currencies of intermediate value

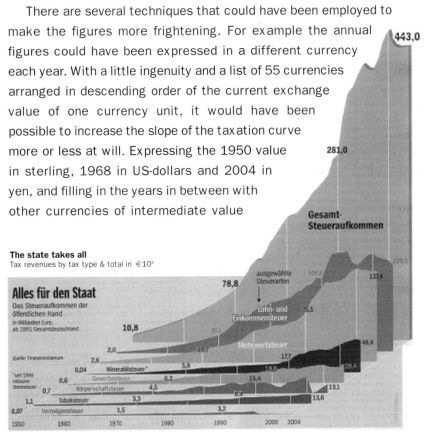

The state takes all
Tax revenues by tax type & total in €10⁹

Der Spiegel, 2005

would certainly have enlivened the chart, though admittedly STD levels would have been stratospheric.

In fact, *Der Spiegel* used a variant of this technique, also with 55 different currencies, but where the most noticeable part of the currencies' names was the same throughout. So the 1950 figure is in 1950 euros*, 1951 in 1951 euros, 1952 in 1952 euros and so on. The result is highly gratifying and is lent a touch of class by showing the contributions of the main different types of tax in front of the total curve. In the face of such ingenious graphical virtuosity, it would be churlish to point out that the finished chart is a meaning-free zone. It differs in no important respect from the apparently absurd proposal described above to use an ordered selection of different national currencies to achieve the same effect.

The PDQ involved in showing a 41-fold rise in total taxation in this period, as opposed to a real rise of just less than 10 times (41/9,56 = 4.3) is just over 4:1, an outstanding value unlikely to be diminished significantly by plausibility problems even in supposedly inflation-conscious Germany. Indeed suspicion of economists is so widespread that nominal values probably enjoy lower STDs than real. They are, after all, what you would have seen on your bank statement or been able to count in your wallet at the time.

To put it another way, substituting a meaningless for a useful figure perversely increases your chances of being believed.

Using number of instances rather than volumes or values

Not surprisingly it's quite difficult to get hold of reliable figures on most sorts of military activity. Our view of the data tends to be obscured by smoke from early morning encounters with firing squads. Alongside gruesome relics from earlier ages, the Tower of London still lovingly preserves the bullet holed chair used in the last execution on the premises, of a German spy called Josef Jakobs, in 1941. But we need not let holes in the data interfere with our production of military statistics, especially if the cause is just and in the public eye.

---

*strictly speaking nominal 1950 deutschmarks translated into euros at the last exchange rate before the abolition of the DM of €1 = DM 1.95583. This is the rate at which all DM obligations were converted on January 1st, 1997. The key point is to use nominal DM values and convert them at the same fixed rate for every year from 1950 to 1996 and nominal euro values thereafter.

The land mine is a simple weapon well suited to manufacture by relatively unsophisticated economies and, unlike dumdum bullets or lances with bamboo shafts which have been outlawed under the laws of war for a century or more, considerably more dangerous to the indigenous civilian population than to the soldiers who tend to lay them. This may explain why it is more difficult to find statistics on land mine production than on the latest fighter aircraft, missiles or battle tanks.

But the authors of *The State of War and Peace Atlas* did not let this discourage them and beat a tactical retreat from using manufacturing volumes or values to simple instances of states in which manufacturing of land mines was known to occur. A pedant might object that Cyprus's output of land mines is probably so much smaller than that of, say, the USA that they scarcely belong on the same chart, let alone deserve the same prominence as here. But, and this is the joy of such statistics, few know for sure and they aren't telling. It is in effect a chart with built-in deniability, in spite of the fact that quantification is suggested by the varying size of the iconic mines. But even this feature could be excused as a trick of perspective.

The lesson for those practising to deceive is that numbers of otherwise unquantified instances are not a bad substitute for meaningful

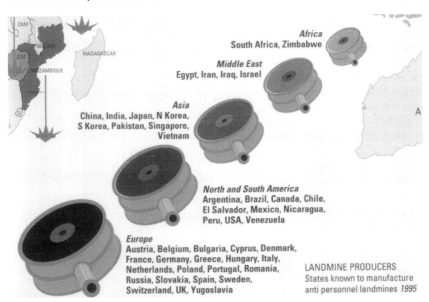

Africa
South Africa, Zimbabwe

Middle East
Egypt, Iran, Iraq, Israel

Asia
China, India, Japan, N Korea, S Korea, Pakistan, Singapore, Vietnam

A

North and South America
Argentina, Brazil, Canada, Chile, El Salvador, Mexico, Nicaragua, Peru, USA, Venezuela

Europe
Austria, Belgium, Bulgaria, Cyprus, Denmark, France, Germany, Greece, Hungary, Italy, Netherlands, Poland, Portugal, Romania, Russia, Slovakia, Spain, Sweden, Switzerland, UK, Yugoslavia

LANDMINE PRODUCERS
States known to manufacture
anti personnel landmines *1995*

Smith, *The State of War and Peace Atlas*, 1997

figures and there is always a reasonable chance that readers' assumptions about the characteristics of the instances, for example that we can ignore differences in scale among them, will subliminally transform a qualitative into a quantitative statement. Sadly, it is not possible to assign PDQ values to this example without accepting the risk of an unacceptably early morning call.

The ten techniques explained in this chapter on manipulating the data before we even draw the first line on a chart give a foretaste of the Aladdin's Cave of graphical deceit possibilities at our disposal. The next chapter concentrates on graphics in the narrower sense and the way we can manipulate the relationship between the data, often already distorted, and the optical impression left by the chart.

4 DISTORTING VALUES

Graphical Manipulation

O nce we've decided on the data, the numerical comparisons that the chart is supposed to reveal, we are ready to use the rich menu of purely graphical devices to adjust that data to the message we want to convey. The one thing that all these techniques have in common is that they lead to charts in which the optical impression. in general or in detail, is not what the reader or viewer would get from the untreated numbers. In other words, the relationship is anything but 1 to 1.

Some of the resulting charts just generally make it harder to derive any message at all from the underlying figures. Such blurring can already be very welcome. But, of course, it is more rewarding to produce a chart that gives a clear, but entirely misleading, message and it is on such techniques that this chapter mainly concentrates.

Disguising shares

Length matters. This is just as true of pie charts, in which lengths have been bent into a circle, as it is of bars and columns, which tend to rely mainly on the lengths of the graphical elements, and of dot and line charts, where the distortion is more likely to take place on one or both of the axes. A lot of the deception techniques, like anxious lovers, try to divert attention from sheer length to even more easily manipulated areas and volumes.

Blurring with πs

Pies are the most popular but also the most vilified of graphical devices. Some authorities even plead for a total ban while others seek to impose curious conditions on their use[1], apparently in the hope of destroying

---

[1] e.g. forbidding the use of absolute values (G.E. Jones, *How to Lie with Charts*, Authors' Choice)

their popularity. But to those who practise to deceive they are a useful part of the tangled web because they have an unrivalled ability to blur comparisons, both of segments within one pie and, even more powerfully, among pies.

We seem to have difficulties with π at the best of times.
Suppose you want, like Puck, to put a girdle round the earth (roughly 25,000 miles) but use a cable three feet too long. How loosely would it fit in terms of the size of the resulting gap between the earth's surface and the cable?

A well-made 2D pie, on its own, leaves little room for manipulation. By the time you've rotated the zero to an unusual position and perhaps fiddled with the segment sequence (both explored under category fraud below) there's not much scope left.

But two or more pies make a graphic deceiver's Christmas, even in two dimensions, because we are poor at judging or comparing angles and even worse at calculating πr^2 (the area of a circle), let alone $4\pi r^3/3$ (the volume of a sphere), in our heads.

The way we produce energy has important implications for us all and is a key topic for ecologists. The figures for 1973 and 2000 can be set out in a simple change diagram. The chart admittedly underlines change rather than absolute values, but it's difficult to imagine any other reason for presenting the data from both years. The graphic effect is clear because we are quite good at judging whether two lines are converging or diverging. The drastic decline in the share of oil is particularly noticeable, not least because it is at the bottom with simple, horizontal base line.

World energy production

Data from: Smith, *State of the World Atlas*, 2003

But if you don't want to admit that our reliance on oil, as opposed to other fuels, has declined in the last quarter of a century, this is a development you may wish to obscure. This is most easily, and fairly unobtrusively, done with a pair of pies. Notice the difficulty of comparing even the 'coal'

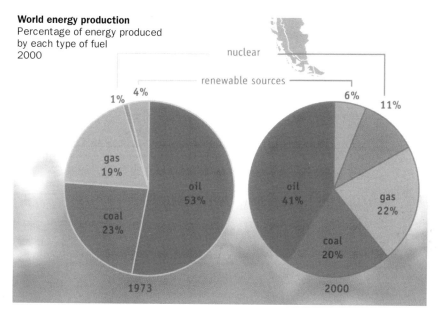

World energy production
Percentage of energy produced
by each type of fuel
2000

nuclear

renewable sources

1% 4%

6%

11%

gas
19%

oil
53%

oil
41%

gas
22%

coal
23%

coal
20%

1973

2000

Smith, *State of the World Atlas*, 2003

segments. A charming refinement of the blurring technique is the unnecessary opposition (clockwise/anti-clockwise) of category order in the left and right pies, which destroys the last chance of graphical comparison between all but the first and last values.

This technique obscures rather than distorting directly and is so frequently seen that its STD (sore thumb discount) values are generally low. Only a very modest percentage of readers assumes that the comparison of segments among more than one pie is a sign that graphical mischief is afoot.

Comparing sizes of whole pies

Yet the chart of world energy production chart does miss an important opportunity to deceive. In spite of the fact that the amount of energy produced in 2000 was very much larger than in 1973, the two pies are the same size. Comparing segments of pies of different size is so difficult that it is an important deception method in its own right. Another chart from the same book gives a good example of just how difficult. The 12% new HIV infections in South and Southeast Asia and the 21% on Eastern and Central Asia look curiously similar.

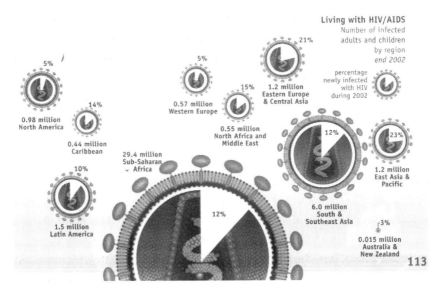

Smith, *State of the World Atlas*, 2003

This technique is far from new. Although gold sovereigns are strictly speaking icons rather than pies, the same difficulties of interpretation can be achieved and were already popular at the beginning of the last century. Compare for example the £21m for machinery with the £123m for textiles, almost exactly six times as much. The area of the textile

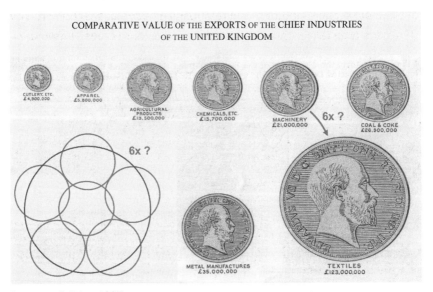

Harmsworth Atlas, 1907

sovereign is also six times the machinery sovereign, the blue circle six times the red, but it just doesn't quite look it. In both charts the area of the pies or icons is roughly proportional to the values portrayed. As our eyes do not give full value to area as opposed to diameter (which is less than three times as great in both cases), both charts are visually understating the size differences: PDQs (potential deceit quotients) 1.3–1.7:1.

Length/area/volume ambiguity

The area of a rectangle is its height times its width; the volume of a cube its height times its width times its depth. So when we are invited to compare the size of two or more cubes, it's rather important to know whether they have been drawn proportional in height, in area or in volume to the data they represent. In fact, we are seldom told how the elements have been scaled and most chart designers seem to feel that graphical convention makes it sufficiently clear. Gloriously, nothing could be further from the truth.

About a century ago there seems to have been a change in the way numbers were visualised. In the 19th century it was assumed that readers would be able and willing to compare areas of graphical elements rather than length. As Edwardians were also much more conscious than we are of the metaphorical nature of the relationship between chart and reality, this made area rather than length comparisons more or less unavoidable

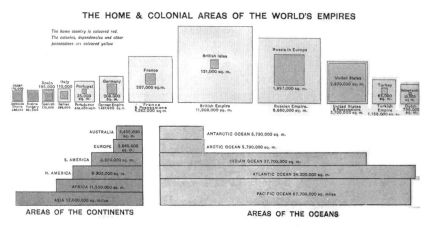

THE HOME & COLONIAL AREAS OF THE WORLD'S EMPIRES

AREAS OF THE CONTINENTS   AREAS OF THE OCEANS

Harmsworth Atlas, 1907

if the values to be compared were themselves areas. The land areas of the various empires and their relationship to the area of the metropolitan power were a source of constant fascination and always used area proportional graphics, as in this example from the *Harmsworth Atlas and Gazetteer* published in 1907. The way the area of the British Isles was dwarfed by the area of the Empire was a source of pride to Edwardians and a thorn in the eye of less well-endowed Continentals.

But it is also noticeable that the area convention was no longer universally applied, even on this chart. The areas of the continents and oceans shown in the lower half of the chart are, with the exception of the Pacific Ocean, also length proportional as the width of the bars has been standardised. By the middle of the 20th century area charts, even for the display of square miles, had almost disappeared and been replaced by charts using length only.

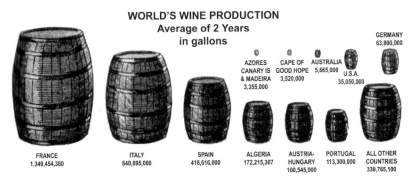

WORLD'S WINE PRODUCTION
Average of 2 Years
in gallons

GERMANY
63,800,000

AZORES
CANARY IS
& MADEIRA
3,355,000

CAPE OF
GOOD HOPE
3,520,000

AUSTRALIA
5,665,000

U.S.A.
35,050,000

FRANCE
1,349,454,380

ITALY
840,895,000

SPAIN
418,616,000

ALGERIA
172,215,307

AUSTRIA-
HUNGARY
160,545,000

PORTUGAL
113,300,000

ALL OTHER
COUNTRIES
339,765,100

Harmsworth Atlas, 1907

Curiously the Edwardians were even happy to be asked to compare the areas of what were obviously 3D objects, in other words, of icons (essentially pictures or visual metaphors – often simplified – instead of bars, columns, dots or lines). In this chart, also from the *Harmsworth Atlas*, although the objects are very clearly 3D barrels, it is their area not their virtual volume that is proportional to the production figures. Whereas we nowadays expect abstract graphical elements like bars and columns to be strictly proportional to their data in length (but of equal width and depth and so also area and volume proportional to the data) there has been no standardisation whatsoever in the last hundred years of the scaling conventions for 3D icons. This means that there is no restriction, beyond the need to maintain a bare minimum of plausibility, on our choice of scale option. We can use icon length, area or volume with complete impunity and there is not even any expectation on the part of the average reader to be told which convention we have chosen. This is not just giving alcoholics the key to the wine cellar; it's locking them in the cellar with a barrel tap and mallet. The ambiguity promises PDQs up to about 30:1 and without significant STDs to restrain ambitions, results can be spectacular.

Using the icon area illusion

Whereas pies tend to blur relationships between values and so help only to minimise data differences, we can use icons to bend the data in any way we want, using the area illusion to change the relationship between graphic and data.

Icons crop up in two uses. Least promising on the deceit scale are large numbers of little icons, stacked or lined up to replace columns or bars. As long as even the smaller values use quite a large number of icons, there's not much scope for manipulation. There's a bit of wiggle room in tampering with the size of the individual icon, which is usually too obvious for serious deceit, but otherwise unintentional humour is the most you can expect from multiple icons.

Single icons on the other hand can seriously facilitate deceit, whether you use one icon divided into parts (to replace the traditional pie) or multiple icons (one for each data category).

In early 2005 the price of petrol at the pump was 82 pence per litre, which broke down into 16p for the crude and refining, 1p for delivery, 5p for

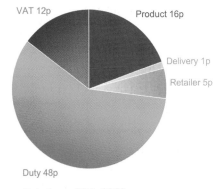

How one litre of petrol breaks down

VAT 12p
Product 16p
Delivery 1p
Retailer 5p
Duty 48p

Data from: BBC, 2005

65

the retailer, 48p for duty and 12p for VAT. Some use these figures to underline the iniquity of a government that pockets three-quarters of the pump price as duty or VAT, while the more sophisticated concentrate on the fact that less than a fifth is proportional to the price of crude oil and that we are therefore less vulnerable to the world's warring factions than commonly supposed. A simple pie chart can cater to both tastes, with tax at one end and the relatively small product cost at the other.

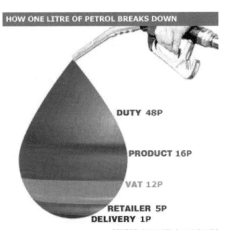

HOW ONE LITRE OF PETROL BREAKS DOWN

DUTY 48P

PRODUCT 16P

VAT 12P

RETAILER 5P
DELIVERY 1P

SOURCE: Automobile Association (AA

BBC, 2005

The more interesting challenge is to exaggerate product cost and minimise duty, thus making the Chancellor of the Exchequer less, and Bin Laden more, frightening. A graphic by the BBC, shown here on the right, goes some way to achieving both these goals with admirable simplicity. The drawn height of each category is proportional to its value – so far, so ordinary – but the teardrop shape eliminates almost half the area of 'duty', and 'product' is exaggerated by being located at the drop's bulging waist.

Comparison with a straightforward divided bar (left) shows the power of the BBC approach, especially in minimising the importance of duty. You can work similar wheezes with almost any transparent container from mere surface tension (as here) to cocktail glasses, bulbous decanters, tilted bottles or balloons. PDQs of this technique reach a respect-

Duty 48p

Product 16p

VAT 12p

Retailer 5p

Delivery 1p

Total at pump   82p per litre

Data from: BBC, 2005

able 2 or even 2.5:1 and some sensible distraction, provided here by the picture of a pump nozzle ('price at the pump', geddit?), should keep STDs below 30%.

Nonetheless, icons only really come into their own when used instead of bars or columns for comparison purposes. PDQs in these cases are limited only by resolution, whether of the output medium or of the perpetrator.

In 2000, the last full year before the terrorist attacks of '9/11', tourism was booming and the ten top tourist destinations worldwide attracted a total of 352 million visitors. France was in first place with 75, the USA in second with 51 and the UK in sixth with 25. The visitors attracted by each country in the group can be shown with a simple bar chart and a restricted grid helps to make clear that France attracts 3 to America's 2 and the UK's single visitor.

Tourist destinations

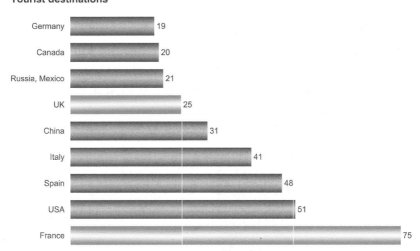

Data from Smith, *State of the World Atlas*, 2003

But, of course, there are those who feel that this is insufficiently flattering both to the Eiffel Tower and to the Statue of Liberty and also, perhaps, that a simple bar chart scarcely does justice to the full romance of tourism. Multiple icons can help to overcome both limitations.

In the example overleaf, also from the *State of the World Atlas*, the wingspans have been drawn in proportion to the number of tourists. What could be fairer than that?

The trick lies in the area illusion. The 'France' icon seems more than three times as big as the 'UK' (its area is in fact $3^2 = 9$ times as big) and there is probably a faint whiff of volume illusion (the volume portrayed is $3^3 = 27$ times as big). However the volume effect is weak. It's weakened further by the perspective feeling generated by the arrangement of the aircraft icons into an approach path. So the effective PDQ is closer to 3 or 4:1 than the theoretically possible 9:1.

However, the winding approach path does brilliantly destroy any remaining impression of a left-hand, zeroed axis. This allows the area illusion to work its full magic, untrammelled by pedants ogling the right hand extremities of the wing tips as they can with the right extremities of the bars above. It also confuses the eye sufficiently to raise the STD to something close to unity.

Destinations
Countries receiving the highest
number of tourists 2000

Germany
19 million

Canada
20 million

Russia, Mexico
21 million

UK
25 million

China
31 million

Italy
41 million

Spain
48 million

USA
51 million

France
75 million

Smith, *State of the World Atlas*, 2003

Difficulties in overcoming the icon area illusion

A special strength of the icon area illusion often gets overlooked. No matter how pure your motives, the moment you reach for an icon, the illusion is simply almost impossible to avoid.

The US Bureau of the Census is one of the most experienced and fastidious producers of statistics in the world and has clearly trained its employees to view the icon area illusion much as Jesuits view a Black Mass or Orthodox Rabbis pork scratchings. So when one such employee

**Times per Week 3–5 Year Olds are Read to by
Race/Ethnic Group* of Parent and Poverty Status**
(Percent of children read to by their parents)

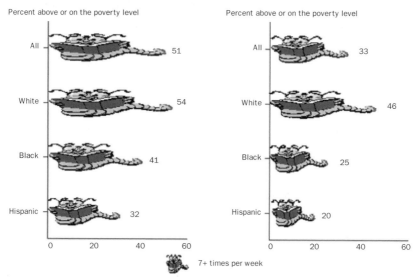

*White and Black races exclude people of Hispanic origin.

US Bureau of the Census, *Survey of Income and Program Participation*, 1998

was taken up to a high place and tempted with all the glories of the graphical world to enliven her chart of how many 3–5 year-olds were encouraged by bedtime reading to become 'bookworms' themselves, she weakened but did not fall. The result was a startling chart, plastered not merely with multi-coloured bookworms, but with multi-coloured bookworms whose width had been distorted beyond human recognition in the attempt to avoid the icon area illusion. The attempt succeeded – the chart is free of graphic deceit – but legibility and aesthetics were terminally prejudiced.

Another American example, from *USA Today*, also plumbs the depths of comicality in cynical European eyes. Having chosen to represent the main countries of origin of emigrants to the USA down the centuries with icons, *USA Today* had to choose between one or two dimensions for adjusting icon size. The latter would of course have involved graphical distortion of the underlying data. But the former would have led to the median Irish figure being flanked either by an emaciated German on one side or an almost perfectly spherical Englishman on the other. The preferred solution was novel and involved the Irish and English figures either dropping to their knees in awe, or alternatively attempting successfully to sink through the floor in shame, at what they had helped to create.

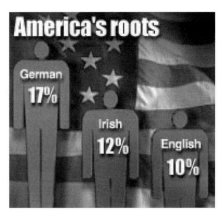

The area illusion (whether using pies or icons) is powerful, with PDQs in excess of 5:1. The closely related volume illusion –

USA Today, 2001

barrels of oil are the crack cocaine of volume fraud – can achieve PDQs well over 20:1 in able hands.

The lesson for graphic deceivers is that all users of individual icons (as opposed to stacks of multiple icons which have problems of their own) have to choose between being deceitful or grotesque. There is no third way, but this may be a blessing in disguise. All who have ever struggled to produce icon-enriched graphics themselves ('just jazz it up a bit will you, dear?') will tend to ascribe that purest of motives, sheer graphical desperation, to its perpetrators. Graphical juries tend to understand this. Area and volume fraud conviction rates are gratifyingly low.

Axis and element interruption

Breaking the axis

If you have ever produced your own graphics, you've probably at some point had to break the value axis to get the data to fit on the page. This happy fact gives comfort and, more to the point, camouflage to those of us who take a more strategic and robust approach to data visualisation.

From the strategic perspective, breaking the axis is usually a way of exaggerating differences between data points. Like first-past-the-post voting systems, as opposed to proportional representation, it tends to sacrifice logic and fairness to clarity and drama. So it is fitting to start with a highly political example: tax as a share of GDP. The raw figures from the first Thatcher government elected in 1979 to the third Blair administration starting in 2005 and projected to 2009/10 can be

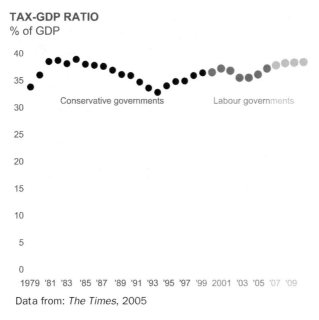

TAX-GDP RATIO
% of GDP

Data from: *The Times*, 2005

plotted as a dotted curve. Frankly not the sort of picture likely to get ministers and civil servants sprinting down Whitehall one step ahead of enraged taxpayers with dripping knives.

The problem of course is the pedantic honesty of the value axis, which starts at zero and increases in equal steps to 40%. Some more excitable commentators on the political right trace the relatively poor showing of the Conservatives in the 2005 General Election to a failure to make clear to the electorate that a change of government would alter the tax figures so dramatically that the difference would even show up on a neutral chart. The use of different symbols to highlight the fact that figures after 2005 are merely projections also limply diminishes the nightmare potential of the graphic, turning it from Ed Munch's 'Scream' into Rodin's 'The Thinker'.

The Times tackled both these problems successfully in a chart of considerable grapho-political subtlety that combines several techniques. The value axis now runs only from 32% to 40%, a reduction of four-fifths, and the power of the now Himalayan curve has been further enhanced by only shading below it. The speculative nature of the last five data points has been marked sufficiently with a thin line and a mal-orientated label to cover *The Times*' small, four-legged beast of burden without detracting

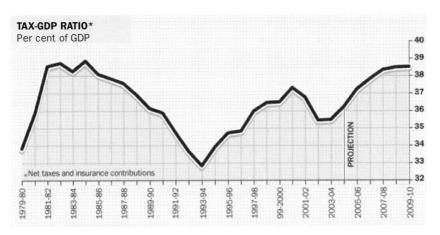

TAX-GDP RATIO*
Per cent of GDP

*Net taxes and insurance contributions

PROJECTION

The Times, 2005

from the suggestiveness of the percentage line itself. The lack of useable grid lines above the curve and of reference points showing changes in the political colour of the government makes it relatively difficult to compare, say, tax levels in Thatcher's first seven years in power with past and future Labour levels. Differences are, so to speak, blaired out of recognition. All in all this is a graphical performance that commands respect and invites emulation, although it is difficult to assign a precise PDQ. The chart gives a first glance impression that taxation has risen enormously since some time around Blair's move to Downing Street and is set to do so again. The PDQ must be somewhere in the range 3 or 4:1 though a good 50% STD applies, leaving us with a still useful 1.5–2.0:1.

Interrupting the graphic elements themselves

These days many charts dispense with scale lines altogether, preferring instead to label the graphic elements directly. As this also lessens the need for grid lines, it saves printing ink and increases the scope for distortion if we are moved to interrupt the graphic elements.

The PR department of German Federal Railways has been suffering for years from the really quite unpleasant criticism that is generated when insufficiently agile road vehicles get deconstructed by fast trains at level crossings. Public hysteria reached such a pitch that the Federal Railways embarked on a programme to reduce the number of level crossings it owned and operated, sometimes even by replacing them with ruinously expensive bridges or tunnels. In 2004 the PR department was given the task of revealing the progress of this humane and far-sighted programme.

The figures themselves were of little help, suggesting that the programme was even more far-sighted than its critics had realised. At the rate of progress evidenced between 1995 and 2003, the last level crossing would be decommissioned around the middle of the century. Even this was optimistic, as a significant proportion of level crossings is owned and operated by a large number of private companies, with little motivation to replace them as the chances of

Number of level crossings on German railways

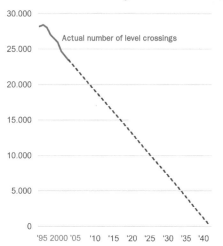

Data from: *Deutsche Bahn AG*, 2006

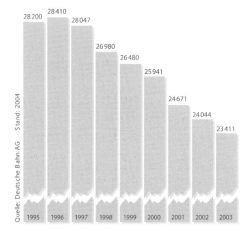

Deutsche Bahn AG, 2006

accidents on any one company's tracks are vanishingly small. But for the press and public a railway is a railway and the Federal Railways are routinely blamed for the rare accidents that occur on private lines between, say, logging camps in rural Bavaria.

Who therefore can reasonably blame the Federal Railways for seeking to put their own efforts in perspective, even by the use of a gradient-improving break in all the columns in the chart? The PDQ involved in showing an actual decline of 17% in the number of level crossings by a reduction in column height of 59% is 59/17 = 3.5:1. The STD value is more difficult to estimate as the Federal Railways incautiously broke every single column in rather a noticeable and elaborately honest way. But even applying an ungenerous 60% discount still leaves a net quotient of around 1.5:1, enough to relieve public anxiety for several years to come.

Interrupting the graphic elements and the scale

Broken axes nearly always dramatise, but given the necessary graphical panache they sometimes have a role in minimising differences too.

Take, for example, the differences in risk, per mile travelled, of using various forms of transport. Using rail or air travel as a reference point, both with a value of 1.0, car journeys are over nine times as risky at 9.1, bicycles 86 times and motorbikes – not for nothing are bikers known as 'organ donors' – are 388 times as risky as rail. Only buses and coaches at 0.9 are less risky.

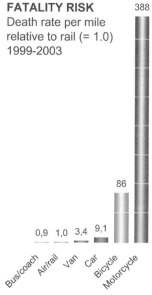

FATALITY RISK
Death rate per mile
relative to rail (= 1.0)
1999-2003

388

86

0,9 1,0 3,4 9,1

Bus/coach Air/rail Van Car Bicycle Motorcycle

Data from: *The Times*, 2005

For most people, the only surprising feature of these figures is the fact that buses and coaches are slightly less danger-ous than aeroplanes or trains. One can imagine an entirely different chart, probably without the figures for two-wheelers, to underline the point. But as the figures for two-wheelers are there, they do completely overshadow all other relationships both in real life and in the graphic.

The one fly in the statistical ointment

goes unremarked: how do we normally think about travel risk? Most people tend to think, if they think about it at all, in terms of 'Am I going to get there alive?', in other words they are interested in risk per trip rather than risk per passenger kilometre. Given that the average trip length in cars could easily be more than ten times that on bicycles, and both are undoubtedly vastly less than that in aeroplanes, the whole chart may need recalculation before it even begins to tell us what we really don't want to know.

But quite apart from such weaknesses in the data themselves, which could be exploited to minimise the dramatic differences portrayed so far, there is also scope for graphical minimisation with the help of a broken axis. Once again *The Times* leads the way.

The key is not to show a break in the left-hand column for the relative death rate for motorcyclists. Putting a squiggle in the scale between 55 and 186 is adequate posterior protection and avoids spoiling the visual impression that motorbikes are only about half as dangerous again as bicycles.

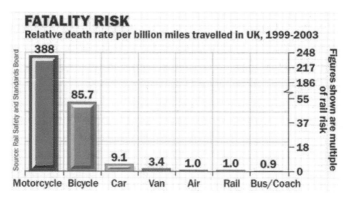

The Times, 2005

What lifts this graphic out of the common rut is the way that it uses two scales, one explicitly on the right, the other implicit in the figures above the columns. The weird choice of scale intervals on the right, and indeed the different scales used below and above the break, remove any justification for the scale except as a hook on which to hang the scale break. The result is a brilliant, impenetrable chart.

Disguised length

So far we have been looking at indirect methods of graphical deceit. These either shift our attention from lengths towards areas, which may have a different relationship to the underlying data, or change apparent data values by manipulating the value scale. This section looks at ways of changing apparent values more directly by working on the chart elements themselves. There are four main ways of doing this:

1 Using 3D and perspective to blur or distort those chart features that show value (for example the ends of bars and columns or the angles between segment in pies).

2 Using other methods to defocus bar and column lengths

3 Adding or subtracting fixed length elements (usually icons) to or from columns and bars to minimize or dramatize differences between them.

4 Using nearby chart junk as a conjuror uses gesture to misdirect our attention

Disguised length – using 2½D to blur value

During 2004 *The Times* reported differences of opinion between the Chancellor of the Exchequer and the City, as represented by the major financial institutions, about the rate at which the British economy would grow during the year. A year later similar disagreements arose over growth prospects in 2005 and 2006. The neutral version of these figures is given below. Journalistically, the bigger the disagreement the better but, atypically, *The Times* dithered graphically.

On November 2003 the disagreement amounted to a mere 0.65 percentage points, scarcely the stuff of which sovereign debt default is made, but good enough for a few column inches with graphical help. The cynical layman is probably more impressed by the failure of both Brown and the City to get it right only a couple of weeks before the end of the year in December 2004. So making the most of this disagreement required breaking the value axis (without being too graphically obvious about it). Curiously, despite breaking the axis, *The Times* also went fully 3D (as opposed to shading suggestive of solidity but without perspective

BROWN v THE CITY

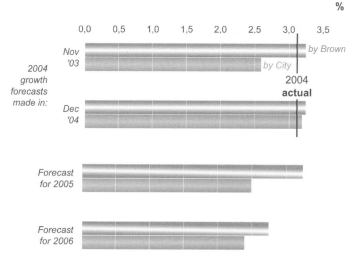

Data from: *The Times*, 2005

effects as in the neutral version above) thus counter-balancing the effect of the broken axis by blurring the differences in the height of the columns. The very faint and misaligned graph paper background also hinders comparison. Note how even the lowest, 2.0%, level is curiously insubstantial. Between the two counter-vailing effects something not all that far from honesty is achieved, possibly inadver-tently, though the main question – is there going to be a repeat in 2005 and 2006? – is obscured by dividing the chart.

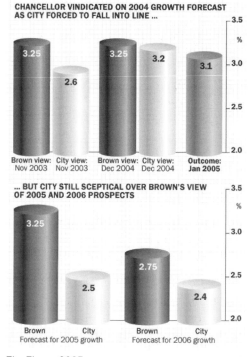

The Times, 2005

As this example shows, you can't get all that much blurring out of using a simple line of 3D columns. It's difficult to push PDQs much beyond 0.8 to 1.2:1, which is still unexciting, even when combined with STDs of almost 1.0.

Disguised length – using 3D to blur value

The real conjuror's hat, Dr. Caligari's Cabinet or Bermuda Triangle of quantitative data is the full perspective 3D chart. The US Bureau of the Census goes to great lengths to check the accuracy of its figures, in this example the number of housing units per county, and to identify factors that tend to increase the error margin, the vertical axis in this chart. The slightly unexpected conclusion is that the accuracy of Census Bureau housing figures is worst in counties with few houses (the front of the depth axis) but, more obviously, whose population is growing or shrinking unusually fast (left or right extreme of the horizontal axis).

Accuracy of 2000 County Housing Unit Estimates by Size and Growth Class

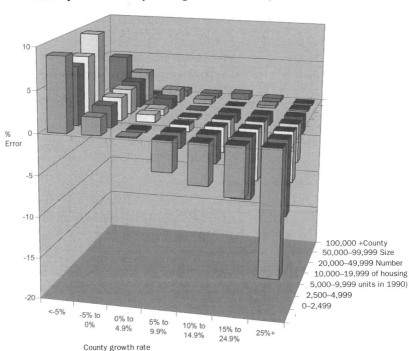

US Bureau of the Census, *Evaluation of the State and County Housing Estimates*, 2003

From the point of view of deceit potential there are several lessons to be learnt from this almost embarrassingly honest example.

First, if you have a needle to hide, a full perspective 3D bar chart offers a gigantic haystack. Quite a lot of the figures for medium-sized counties with moderate growth rates, those in the middle of the array, have simply disappeared behind others.

Second, most of what is not hidden is gloriously blurred. Even for the fully exposed columns, it is impossible, without using a ruler, to determine their actual length by reference to the grid. How long, for example, is the longest bar in the array (front row right) not to mention the ones behind it?

Finally, if you let your gaze linger on this graphic for more than a second or two, the chart will suddenly spring to life and reverse its background. This is the Necker Cube illusion at work, of which a simple version is shown in outline. Is the circle on the back or front wall? What you first see in the Census example as the view of a dark grey floor from above, becomes the bottom of a tree house seen from below and the most distant left hand corner suddenly projects like a mediaeval jakes sticking out over a foetid castle moat. This also reverses the order of the scale for the number of housing units and seems to put the most populous counties at the front. Meanwhile the bars themselves do a René Magritte and hang, like tiles falling off a roof, in mid-air in front of

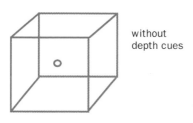

without
depth cues

with depth cues

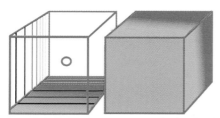

The Necker Cube flips. The nearest wall always seems smaller: a false perspective effect. The flipping itself has more to do with the lack of visual data to distinguish interior and exterior corners. The visual cortex – or a pre-processor between retina and brain – seems to treat such corners as quanta, irreducibly simple components or 'geons' rather than, more analytically, as the convergence of three lines.

The effect can be diminished by providing depth cues but is still difficult to eradicate.

the tree house. The chance of anyone wasting much time on the actual figures during these aerobatics is negligible.

3D bar and column charts confuse and repel more than they distort. But getting some of the data to disappear altogether and making 80% of the rest unusable must be worth an honorary PDQ of 2 or 3:1 any day, with almost no discount. The thought of so powerful a weapon getting into the hands of a less scrupulous organisation than the Bureau of the Census is deeply moving for those of us interested in graphic deceit.

In the meantime, we can enjoy the uses to which artists have put perspective illusions in general and the Necker Cube in particular.

 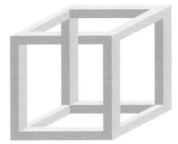

William Hogarth, 1754

Disguised length – other ways of blurring length

Not all blurring of the length of graphical elements has to be done with 3D. With careful graphical treatment it can sometimes be difficult to decipher the value behind even 2D bars or columns.

The best techniques simultaneously attack as many of the elements that define length as possible. Even fairly simple columns tend to define their own height in a surprisingly large number of different and redundant ways. In this example, the height of the column ABCD is defined in no less than seven different ways. With the possible exception of the data

1 Length of line AD

2 Length of line BC

3 Height of shading inside ABCD

4 Height of line AB above zero

5 Area of line figure ABCD

6 Area of line figure ABCD

7 Number label '100'

label '100' itself (which should be omitted altogether in the interests of distortion), this redundancy opens the barn door to creative blurring.

The key is line AB at the top of the column. It is so definite and relates far to clearly to a possible value scale (not shown here). Changing it into a diagonal line, with B higher up the scale than A, not only makes it unclear whether A or B, or perhaps an average of the two, defines the value but also creates the same ambiguity in the relationship between lines AD and BC. It makes the size of the shaded area and of line figure ABCD equally obscure. Elements now contradict each other and comparisons among two or more such mutilated bars border on the meaningless.

Of course, as in even the best conjuring tricks, it is difficult to imagine anyone falling for it when the individual moves are revealed in slow motion and when the whole is presented on an antiseptic white background. Peering over Leonard da Vinci's shoulder while he daubed away would probably transform even the Mona Lisa's famous smile into a George W. Bushian smirk. So it is reassuring to see a real life example, from the *Atlas of Human Sexual Behaviour*, which has used precisely these manoeuvres to subvert one of

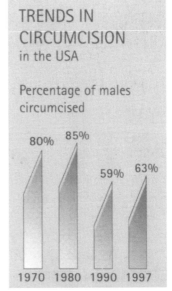

TRENDS IN CIRCUMCISION in the USA

Percentage of males circumcised

80%   85%   59%   63%

1970 1980 1990 1997

Mackay, *Atlas of Human Sexual Behavior*, 2000

the basic principles of graphic display, even if in this particular case it does leave half the audience nervously clearing its throat and crossing its legs.

Disguised length – add or subtract constant graphical elements

Part of the, limited, charm of the last example lies in a further graphical trick: subtraction of a constant from each column. Because the length of the cutting edge is the same in each column a triangular figure or geon, of constant height has been removed, perhaps more exactly half removed, from every column. But the eye gives greater emphasis to the full width portion of the column. The tips of the scalpel blades for 80% and 59%, a difference of almost exactly a quarter, are indeed shown as 25% apart. But because a constant has been subtracted from both, the full width portion of the 59% is less than half as high as that of the 80%, dramatizing the difference.

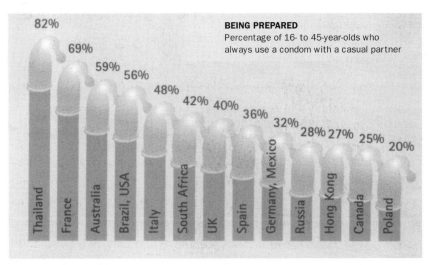

Mackay, *Atlas of Human Sexual Behavior*, 2000

An even more striking version of this effect, in which it takes clear precedence over a rather anaemic version of the ill-defined-end-of-the-bar technique, occurs some pages earlier in the same book. As the value labels make clear, the Thais are just over four times as prepared as the Poles, 82%:20%. Yet the Thai column is only two-and-ahalf times as high as the Polish. On the other hand, because a condom icon obscures a constant length of each column, the uncondomed remainder below it exaggerates the difference, putting the Thais at five-and-a-half times the Poles. Overlooking the sore thumb discount applicable to a chart that manages to get it wrong twice in different directions, this is a good example of the power of the constant subtraction method with a PDQ in this example either of 1.6:1 or of 0.7:1 and an STD of 0.8–0.9.

Unnecessary graphical honesty can sometimes drive a determined deceiver almost to drink. In blatant disregard of the manipulative possibilities of subtracting constants from graphical elements, USA Today went to the trouble in this example of adding rubbers, sorry, erasers, to both pencils but then made the more conspicuous pencil shafts, not the whole of the pencil, proportional to the data. However something was

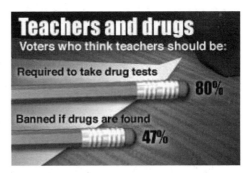

USA Today, 1998

saved from the wreck. The resulting chart leaves us not quite knowing which way to look. Its ambiguity about whether a constant was initially added and then ignored or added after the pencils were already correctly aligned provides a suitable bridge to the next example.

The same constant subtraction concept can be used in reverse, as constant addition, to understate differentials. The chart overleaf purports to show the number of successful launches into earth orbit since the first Russian Sputnik in 1957, but seems to have difficulties on the category axis distinguishing French and European and Soviet and Russian launches. It even includes the one and only British satellite launch, the Prospero X-3 launched from Woomera in October 1971 to avoid the postage and packing on returning spares to the UK when the Black Arrow programme was cancelled.

By adding a nose cone and escape module of the same length to the top of every rocket, the chart achieves a column for 2,289 Soviet launches that is only five-and-a-half times as long as the column for the one UK launch, a PDQ of a mindboggling 2,289/5.5 = 416:1. Indeed it

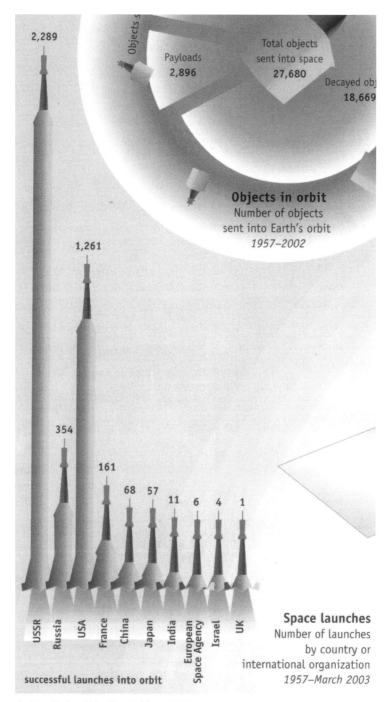

2,289

Objects s

Payloads
2,896

Total objects
sent into space
27,680

Decayed obj
18,669

Objects in orbit
Number of objects
sent into Earth's orbit
1957–2002

1,261

354

161

68 57

11 6 4 1

USSR
Russia
USA
France
China
Japan
India
European
Space Agency
Israel
UK

successful launches into orbit

Space launches
Number of launches
by country or
international organization
1957–March 2003

Smith, *State of the World Atlas*, 2003

arguably reaches a PDQ of ∞:1 in showing 1, 4 and 6 (for the European Space Agency, Israel and the UK respectively) as equal. Clearly a sore thumb discount well over 90% must be applied, but less exuberant variants of this technique are still powerful and can go unnoticed and so undiscounted.

Disguised length – manipulating perspective on 3D pies

The exploitable difficulties in comparing segments among more than one pie chart and even more in comparing the size of whole pies have already been discussed. But like columns and bars, pies are also intriguingly vulnerable to determined distortion by 3D perspective. This is easily remembered as the 'bouncing aspirin' technique.

Take the question of what the government spends our money on. Total government expenditure in the UK in 2004 was £519bn of which social protection, health and education accounted for almost 60%. The neutral chart below shows the proportions of total expenditure by area and seems to be quite hard to distort. But suppose you are a farmer, living in subsidised housing in a remote area. You might want to play down the amount that government spends on transport, agriculture and housing when making your case for redistributing the budget. In this case, the bouncing aspirin could cure more than headaches.

Total managed expenditure: £519bn

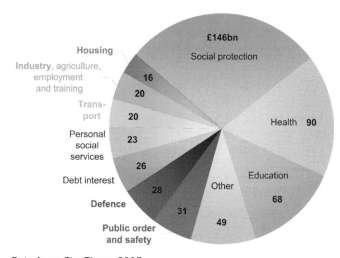

Data from: *The Times*, 2005

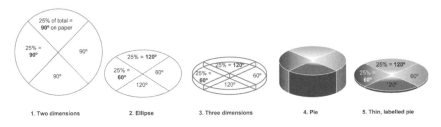

1. Two dimensions   2. Ellipse   3. Three dimensions   4. Pie   5. Thin, labelled pie

The technique uses the fact that the aspirin has to turn the round 2D pie into a 3D-suggesting oval or ellipse. All pie segments out at the higher curvature parts of the ellipse, in this case to the left and right, are minimized compared to segments along the stretched foreground and background of the pie. Increasing the thickness of the aspirin can further heighten this effect. This, like a well-advertised hair shampoo, gives added body to the segments at the front, the only place you can see the thickness, at the expense of the rest.

Compare the apparent importance of housing, industry and agriculture and transport, in total £56bn, with the similar total for defence and public order and safety at £59bn. In the neutral version the defence and police share does look very much the same size as the housing, industry and agriculture and transport share. But in the bouncing aspirin, guns loom considerably larger than butter and buses even though our brains do compensate slightly by taking some, but by no means all, of the perspective effect into account. It is slightly surprising that a newspaper of this political persuasion allowed this particular government to appear

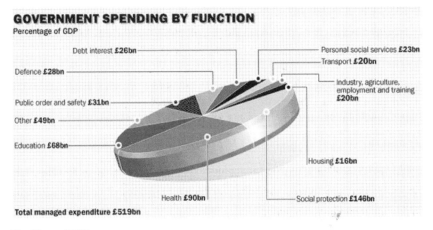

GOVERNMENT SPENDING BY FUNCTION
Percentage of GDP

Debt interest £26bn

Personal social services £23bn

Transport £20bn

Defence £28bn

Industry, agriculture, employment and training £20bn

Public order and safety £31bn

Other £49bn

Education £68bn

Housing £16bn

Health £90bn

Social protection £146bn

Total managed expenditure £519bn

The Times, 2005

to be so keen on defence and law and order relative to other spending departments.

Of course, if you want to distort some values and simply get others to disappear, you can build a column of aspirins sufficiently close together for the hindmost half to be completely obscured, but STDs quickly become exorbitant. To sum up: achievable PDQs approach 2:1 (as shown in the simplified diagram). Tilting the aspirin even further can increase this, but at the price of increasing illegibility. For not entirely obvious reasons, the sore thumb discount is bearable: all in all a worthwhile deceit technique.

Disguised length – chart junk

Sometimes none of these ways of directly manipulating the perceived size of graphical elements seems adequate. When all else fails and it seems that the truth will not only out but go rampaging across screen or page, we can always fall back on chart junk as a fig leaf to cover the inadequacies of our quantitative arguments. This section deals only with the use of chart junk to disguise the length, or otherwise, of graphical elements, other applications are discussed later.

According to UNICEF, the United Nations Children's Fund, First World countries increased their annual GDP per head by almost US$8,000 between 1990 and 1997. During this period their contributions per head to development aid for poorer countries fell by US$18 per head. The intended UNICEF message seems to have left the unfortunate graphic designer having to put $18 and $7,878 on the same scale.

Most layout designers would have weakened and begged to be allowed to use simple text and table along the lines of:

Increase in income per head:	**$7,878**
+ decrease in aid per head:	**+$18**
= gratefully pocketed for expenditure on wine, women and song:	**$7,896**

or at least asked for different units, perhaps share of GDP spent on development aid, which indeed fell precipitately, as the end of the Cold War made overseas development aid (ODA) seem less strategically important.

Instead UNICEF went for chart junk. The first impression is powerful enough and not completely misleading: GNP went up a lot, ODA went down, but a lot less. During a visual presentation, in a darkened room without audience eye contact and a computer powerful enough to change pictures at almost subliminal speeds, some speakers would be prepared to give it a whiz. It's amazing what you can get away with. But putting it in a printed report took guts rather than brains. The lesson is quite simply that there is a limit to what even chart junk can achieve, but also that there are probably other ways of achieving the required distortion.

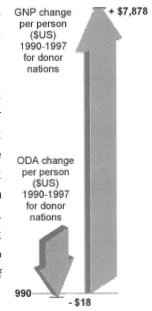

GNP change per person ($US) 1990-1997 for donor nations — + $7,878

ODA change per person ($US) 1990-1997 for donor nations

990 — -$18

UNICEF, *The State of the World's Children*, 2000

Disguised length – last resort: just hide one end

You might think that this chapter has already exhausted the last reserves of human ingenuity in disguising the length of graphical elements. But the very best, the non plus ultra, electronically tested, Columbus's egg of the genre, which makes all other techniques look ponderously pusillanimous, is simply to put something else on top of the element whose length you want to hide. The academic cap on the highest average salary even makes it impossible to calculate a PDQ. We just have no idea of the scale of the obfuscation and must either give up or cobble something together from the dollar figures. On the other hand, elegant as it is, its STD is at least 1.0 and so the effective deceit achieved is zero. Good for a laugh but not really worthy of a place in a serious deceiver's toolkit.

$63,229

Education pays

Average salary by education level

$40,478

$22,895

Advanced degree — Bachelor's degree — High school

USA Today, 1999

Disguising length by manipulating zero point

Move zero base upwards

The easiest, and sometimes the subtlest, way of disguising the length of graphical elements like bars and columns is surreptitiously to shift the zero base on which the elements stand. The results are similar to breaking the axis, with or without warning signs, but are generally more difficult to spot and so are less heavily discounted.

Lots of Yachts
Number of sailing boats in Germany in 000

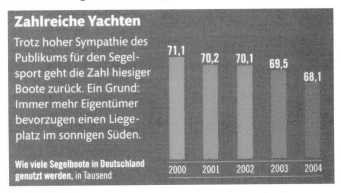

Capital, 2005

Take for example this depiction of the fastest, voluntary shrinkage in the German fleet since the Imperial German Navy was scuttled in Scapa Flow in 1919. Interestingly for students of the power of the media, this seems to have been mainly the result of the German admiral's misunderstanding of British press reporting of the Versailles Peace Conference. In this case however, the numbers are about sailing boats rather than battleships. Instead of interrupting the columns, *Capital*, one of Germany's leading business monthlies, simply shifted the zero base of the value axis upwards. The decline from 71,100 to 68,100 (–4.2%) is indicated by a reduction in column length of 26%: PDQ = 26:4.2 = 6.2:1. As the shifting of the zero base is unlikely to be spotted, the STD is almost zero and so the net distortion is at least 5:1, enough to have High Admiral Reinhard Scheer, the victor – according to German accounts – of the Battle of Jutland, ringing for full revolutions from the grave. The simplicity and elegance of the deceit are almost as admirable as its power.

Examples of this technique are so frequent that STDs have been worn paper thin by simple repetition. As politicians have a healthy instinct for never being ahead of the public on any issue, this may explain why it is particularly popular in Westminster. It is far and away the most commonly used graphical deceit technique in British politics. It has even been spotted at Prime Minister's press conferences at No. 10 in spite of the fact that New Labour, like the Liberal Democrats, has generally been at pains to eschew anything as elitist as numerical arguments in most of its promotional material. Curiously only the Greens and the Conservatives seem to share a taste for numbers and PowerPoint.

At a prime ministerial press conference on 30 July 2003, the head of the Prime Minister's Delivery Unit delivered a PowerPoint presentation of some 60 slides to explain the Unit's purpose and show progress to date. One of the areas covered under the beady eye of the Prime Minister himself was the reading achievements of English 9–10 year olds compared to children from other advanced and less advanced countries in 2001.

Quite apart from lingering scepticism that any government would have had a major impact on reading ability within five years of taking office, the figures themselves are difficult to interpret. First, the differences are just not very big: only 11% separates England, the third best performer, from

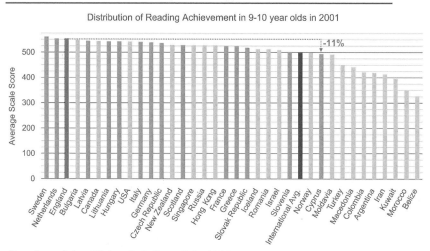

Plateau is frustrating ... but current performance is world class

Distribution of Reading Achievement in 9-10 year olds in 2001

Data from: *Prime Minister's Delivery Unit*, 2003

Cyprus, the worst performer in the European Union. Second, the fact that English 9–10 year olds have already been 20% longer in school than in most other developed nations, but can look forward to 30% less as most leave school early by international standards, also reduces the significance of the English results. Finally the very wide spread of scores within each country means that the rank order among the top dozen is anything but statistically certain. In other words, we're in urgent need of something, anything, to quell these doubts and make it clear beyond a peradventure that England wins at least the bronze and is seriously better than all but two of our Continental neighbours.

No. 10's solution was simply to discard the bottom half of the chart, without so much as a squiggle or footnote to draw attention to the disappearance of the zero base. This rather neatly suggests that England is about 25% above the average of all the countries shown here, whereas the difference in fact is about 10%. A PDQ of 2.5:1 is a very respectable achievement and can expect to go undiminished by any STD at all. This is 'Delivery' indeed, confirmed by the knighthood for the head of the Unit two years later.

A more complex example is provided by a contribution to the Conservative Party 2005 election campaign by the Conservative Policy Unit called The Drivers of Regulation. The chart was based on a Single

Plateau is frustrating ... but current performance is world class

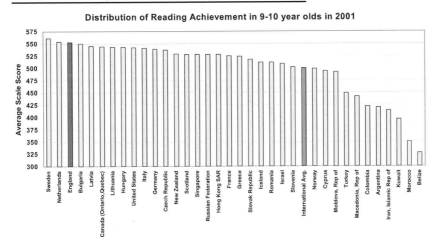

Distribution of Reading Achievement in 9-10 year olds in 2001

Prime Minister's Delivery Unit, 2003

Simplicity of regulatory environment when trading
(4 = simplest, 0 = most complex)

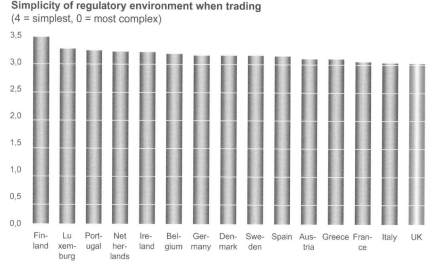

Data from: Conservative Policy Unit, *Drivers of Regulation*, 2004

Market Monitor Survey for the European Commission from 2001, which asked several thousand businessmen in the EU to rate the regulatory environment in other member states when trading within them. The survey results seemed to show that although British business suffered the most, there wasn't much to choose – about 12% – between best and worst, among the then members of the EU when it came to the simplicity, or otherwise, of business regulation. We intuitively expect complexity to be depicted as 'larger' than simplicity but, on the other hand, 'good' values to be higher than 'bad'. The scale and orientation of this chart somewhat awkwardly follow the latter expectation, but the graphic is neutral.

But of course it won't do at all. We all know perfectly well that British business is choking to death in the toils of a bloated Brussels bureaucracy, whose pettifogging regulations are largely ignored on the Continent but gleefully applied to the letter by Downing Street.

Obviously the simplest solution to our graphical problem lies in moving the zero base upwards to focus readers' attention on the differences among states. But, if we were graphically honest, we would then have to make the scale break reasonably obvious, perhaps like this. The more interesting challenge lies in camouflaging such an obvious technique and involves several discrete steps.

Simplicity of regulatory environment when trading
(4 = simplest, 0 = most complex)

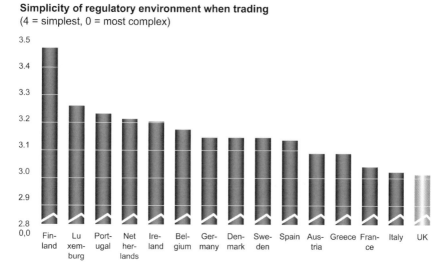

Data from: Conservative Policy Unit, *Drivers of Regulation*, 2004

If we change the title from 'Simplicity of regulatory environment ...' to 'Complexity of regulatory environment ...' not only does the UK now get the longest, rather than the shortest, column but we also get an excuse to make all the columns hang downwards emphasising the negative. So far, so more or less reasonable. But the key trick is to do this *without* inverting the value axis. This means that the hanging columns no longer represent each state's average score but rather the

Complexity of regulatory environment when trading (4 = simplest, 0 = most complex)

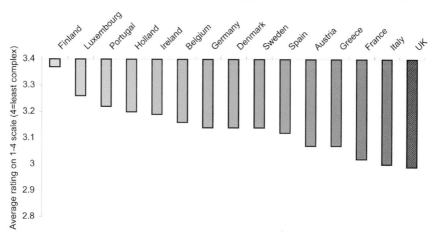

Conservative Policy Unit, *Drivers of Regulation, 2004*

difference between that score and whatever we decide to use as the largest simplicity value. To put it another way, it would show non-scores rather than scores.

As every school child knows, there is no moral or graphical requirement to show a scale break above the largest value shown, almost 3.4 in the case of Finland, nor, in this inverted case, to show a break below the lowest score at, say, 2.8. By graphing non-scores instead of scores we are now free as air and under no obligation whatsoever to clutter up the axis with embarrassing scale breaks. The only limitation is not choosing a maximum value that would make Finland disappear altogether or a minimum that would cut the bottom off the UK column.

This chart shows considerable graphical virtuosity. The PDQ is around 7:1. The Finnish score was some 12% points higher than the UK's, yet graphing non-scores on a broken axis makes the Finnish column 83% shorter than the UK's. Ignoring the sign and just concentrating on the absolute difference gives a PDQ of 83:12 = 6.9:1. Under these circumstances there is no need even to discuss STD values, making this one of the most powerfully transformational charts in this book. Pity about the election, though.

Other changes in the position of the zero base

Surprisingly, it's also possible to shift the zero base explicitly yet still achieve a useful degree of distortion. Let's take the average age at which children receive their first sex education in different countries, having first suppressed the usual doubts about the reliability of such sex survey results. Surveys of sexual behaviour are always at the mercy of translation problems and other cultural specificities and in this particular case the identity of the commissioning organisation, Durex Condoms, could give rise to concern in the exceptionally statistically fastidious. In 1999 the average age at first sex education lay between 11.3 and 13.5 years, a spread of 2.2 years or over 20% of the UK figure.

Suppose we want to play down the precocity of British children's sexual knowledge. This might be simply for reasons of scientific caution, as the figures don't correspond all that well with more objective measures of sexual incompetence like teenage pregnancies. For example, teenage pregnancies are most common in the USA, which only achieves a roughly average position on these sex education figures. How can we minimise the differences between countries graphically, without resorting to the neutron bomb of a logarithmic scale, which leaves charts standing but which massacres meaning?

A regular shifting of the zero line from one category to the next offers one solution, as in this example from the *Penguin Atlas of Sexuality*. It is all the more elegant for using a mid-line as the effective zero. This both camouflages the shift and even offers a jolly, game-playing reason for it. If you let your gaze linger on the chart you can even see how a 3D inter-pretation (in which the German column would be 'further away' than the Thai), which would suggest similar ages of first sexual education in all countries, could emerge. The power of the chart to mislead is, of course, diminished by the explicit column labels, but needing value labels to correct a visual impression is already halfway to winning the battle to deceive.

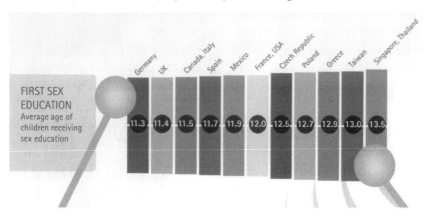

Mackay, *Atlas of Human Sexual Behavior*, 2000

As we have difficulty in adding the top and bottom differences together, this deceit technique has an underlying PDQ of up to 2:1 and an STD close to zero. This is useful, as long as you can find an exculpatory visual metaphor, like a xylophone, to provide the background musak.

Omit zero (or its equivalent) altogether

Not all base lines necessarily have to be explicitly at zero. Under certain circumstances a base of zero is itself misleading and distortion is only eliminated if it starts at one. It is easy to check if the base line has been given a non-distorting value by taking a couple of numbers further up the scale and seeing if the usual relationships have been preserved. For example, is the distance between the grid lines for the base and 2 the same as the distance between 2 and 4. We can think of this as the two-plus-two-equals-four check and it should be born in mind whenever we adjust value axes.

However, attack is the best form of defence and there is no harm in making that comparison as difficult as we can. In this example, *The Times* was reporting research that showed that the risk of certain birth defects rises quite steeply with increasing age of the father. At the age of 45 the risk of fathering a child with birth defects is six times greater than at the age of 25. But what is the risk at the age of 25, two times something, but what? It gradually becomes apparent that the y-axis has been mislabelled and should start at 1 times, in other words where the

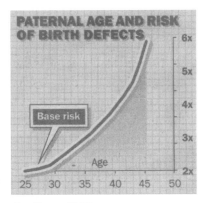

The Times, 2005

supplement to the base risk is zero. If this is corrected the two-plus-two-equals-four check is passed, leaving only the mystery of the gap, unfilled by figures, between the age of 45 and 50, a category into which many fathers, particularly of second or subsequent children, these days increasingly fall. The gap is an optical invitation to extend the line upwards and to the right and extrapolation suggests that it would meet the vertical axis at a risk of 10–12 times the base risk. Or, to put it another way, fathers over 45 are grossly irresponsible.

So, apart from what looks like a typographical error, there is no distortion involved here? In fact, however, it turns out that the graphic designer was given the wrong data from the original Danish research paper, which concluded 'Overall, there were no differences in the prevalence of malformations as a function of paternal age'. Oh dear!

Breaking the axis or the graphic elements, like manipulating the zero base, are powerful techniques of distortion by omission. It's worth considering why the STDs of these techniques are usually so forgiving and their net deceit values therefore so high.

The main reason is that there are some occasions when omission of data in this way does not distort, when it may even improve the clarity and accuracy of the chart. But these occasions are difficult to define and so the highly welcome default position has emerged, that scale and element breaking and zero shifting are generally permissible techniques for accommodating large values and magnifying significant comparisons. Nonetheless it's worth knowing when these techniques do not distort. After all, from our specialised point of view, nothing could be more irritating than pouring graphical effort into a chart only to find that it merely emphasised the truth.

Breaking scales and elements to accommodate large value spreads does not distort when the broken value is peripheral to the message of the chart. In other words, as long as it's just a way of keeping the category axis complete, there's no great harm in breaking a graphical element depicting an unusually large yet uninteresting value.

The other exception to the general rule – 'Don't break scales or elements' – is even more useful. The one thing that breaking the value axis or manipulating the zero base doesn't distort is the relationship between two delta values. For example, to return to Reading Achievement, wherever you cut off the bottoms of the columns, the top of the Turkey column will always look roughly as far below average as the

Plateau is frustrating ... but current performance is world class

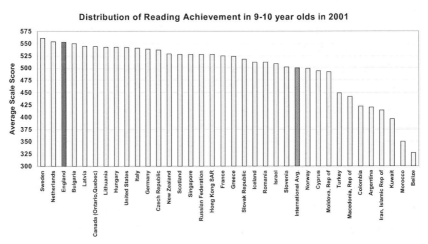

Distribution of Reading Achievement in 9-10 year olds in 2001

Prime Minister's Delivery Unit, 2003

English column seems to be above. So using this chart to make the two-stage comparison from England to International Average to Turkey, shifting the zero base introduces no distortion. To distort this △:△ relationship you'd need to stretch the value axis unevenly, for example by using a logarithmically scaled axis.

So pedants who reject all scale breaks or shifting zero bases out of hand are clearly a rule too far. But we should be grateful to them for discrediting a rule that is usually right and whose consistent application – leading to no scale breaks or shifted zeroes at all – would seriously impair our own freedom of action.

Distracting icons

All charts are metaphors for the real world, whose quantitative relationships they depict. Some graphical practitioners seem to think that the use of icons just moves the metaphor a little closer to reality than the bar, column or pie it replaces. In practice, as we have already seen, icons must either be grossly distorted – and so increasingly unreal – to avoid length/area/volume/distortions or, if left undistorted (i.e. magnified or shrunk in two dimensions), correspond less closely to reality than purely geometric elements that do not have a preferred aspect ratio. This means that using icons to convey values directly is usually punished by high STDs as the viewer realises that something is amiss, either with the icon itself or with its relationship to reality.

But icons do have other roles that are sometimes more useful to those of us wishing to change or cloud perceptions graphically. Apart from grander roles, they can be used to distract from the figures themselves and to invite viewers to let their minds wander freely to other topics.

Unscaled icons

We tend to use icons to make charts and the numbers behind them more attractive to viewers and readers. This idea is well over a century old, as several examples in this book show. But the chart on the right is directly descended from yet older examples, like the sea serpents, man-eating starfish and whales often to be found in empty spaces in old maps and maritime charts. Even the youngest midshipman did not expect to gather useful information about the size of the monsters he would meet in the South Seas from the pictures on the chart. The example on the right carries this idea to its modern graphical conclusion and shows how to break triumphantly free from the shackles of mere scalar verisimilitude.

Mackay, Atlas of Human Sexual Behavior, 2000

Just to take the two icons on the left, the PDQ involved in making the computer (representing 38%) 40% smaller than the couple in the doorway (40%) is around 40:2 = 20:1. Sadly, the STD of at least 1.0 rather spoils the fun.

If you want to take a risk in your attempts to enliven your graphics, make sure that, as here, there is a fall-back justification. If you take the area of the icons rather than their height, the PDQ is quite small as the computer is a lot wider than the doorway and the pager roughly half the area of the computer and so on. But the STD remains stubbornly high and the attractiveness of the chart is not exactly Oscar material.

Irrelevant messages

Whenever you use an icon you are offering a metaphor to your viewers, of which it is all too easy to lose control. It all depends on the viewers and their imagination, which can be frighteningly powerful.

Take for example this chart from a 1920s tract arguing for the return of the German colonies (including present-day Cameroon, Tanzania and Namibia) that were seized by the victors after the First World War and administered until 1945 under League of Nations mandates. The point, such as it is, is to show that Germany had been pouring human resources into its colonies until the beginning of the First World War leading to an increase in the white population from 7,525 in 1902 to 24,380 in 1913. The height of the figures is indeed to scale and the potentially useful

White population of the German colonies 1902–1913

Die Zunahme der weissen Bevölkerung in den deutschen Kolonien 1902/13.										
1902	1903	1906	1907	1908	1909	1910	1911	1912	1913	
Zahl der Weissen: 7525	7788	11 273	12 412	13 858	18 175	20 074	21 667	23 342	24 389	

Sache, *Das Deutsche Kolonialbuch*, 1926

area/volume ambiguity has been cravenly abandoned by cropping the larger figures on both sides. So far, so graphically pious.

But there is a further message based on the choice of each annual icon which suggests considerable sophistication in the designer and which could usefully be emulated today. The figures grow not only in height but also in civilisatory power. The explorer/hunter on the far left only really has his Mannlicher carbine to distinguish him from the locals, the 1903 farmer a rather daring hat, but by 1906–7 the police have arrived, by 1909 a businessman or accountant, 1911 a schoolmaster, 1912 a surveyor and finally, as the apotheosis of the civilising mission, an administrative grade civil servant with a pronounced stoop in 1913. Only the arrival of the slightly Noel Coward figure in the yachting cap and bow tie in 1910 is a little difficult to interpret.

In short, for anyone with imagination, the icons are considerably more entertaining than the data they represent. So if you have data to hide, get a good artist and fail to explain, implicitly or explicitly, your choice of icon.

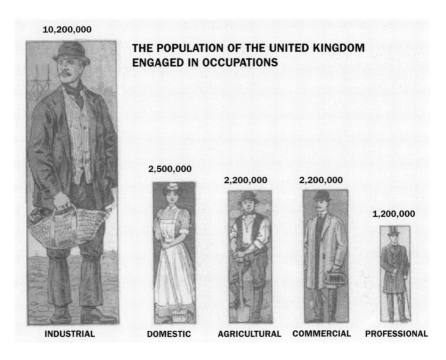

Harmsworth Atlas, 1907

Icons as column filling

Icon choice can be more obvious, and so more boring. In the example on the left, from the *Harmsworth Atlas*, icons are being used simply as column colouring, unless, of course, there were atlas users at the beginning of the 20th century who were so isolated in a life of idle luxury, that they required living legends to identify the members of the main economic and social groups. The distortion, such as it is, derives from the ambiguity about whether we are supposed to to look at the height or the area of the columns, which tends to exaggerate the importance of the professional classes, who probably formed the main market for the atlas. The icons just distract from this mild deceit.

Irregular value axis

As far as the value axis is concerned, we have only looked so far at relatively drastic techniques involving breaks and massive subsidence of the zero point. But the axis also lends itself to more subtle techniques of selective distortion. Foremost among these is the use of the logarithmically scaled axis. This not only lends an air of higher mathematical plausibility to the whole chart but can also be laughed off as an inevitable but unimportant side-effect of the struggle to fit data with large value spans on one piece of paper. Most readers are remarkably tolerant of the latter excuse and so STDs tend to be low.

Using a logarithmic scale

If we are quite honest – a subject on which this book cannot possibly comment – a lot of the scale manipulation described so far can be a bit obvious, which is reflected in STDs. But some forms of squeeze to please achieve their aims particularly subtly. Some of them are so good at simulating an atmosphere of scientific reliability, without any sacrifice in increased clarity, that one is tempted to award STD values below zero. To put it another way, the deceit technique produces a chart that looks more plausible than the neutral version.

The infamous currency and stock market crash in the young tiger economies of Asia happened at the beginning of 1998. The timing of the crisis and the first tender shoots of recovery from it can be read from the rates of the local currencies involved relative to the US$. Within weeks the national currencies of Thailand and Malaysia lost about half their value in terms of the US$.

The International Monetary Fund (IMF) not only reported the changes but also played a major part in supporting the victims economically and financially. It would be ungenerous to criticise the IMF not only for doing good but also for not seeking to diminish its own role in over-coming the crisis. A neutral presentation of the facts would have done full justice to both these aspects. Of course, breaking the axis below US$ index 40 would have

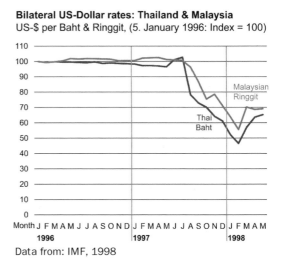

Bilateral US-Dollar rates: Thailand & Malaysia
US-$ per Baht & Ringgit, (5. January 1996: Index = 100)

Month J F M A M J J A S O N D J F M A M J J A S O N D J F M A M
1996 · 1997 · 1998

Data from: IMF, 1998

dramatised things, but in a crudely obvious way unworthy of an international bank like the IMF. The other problem is that it also dramatises the relatively harmless rate changes in 1996 and the beginning of 1997. So the cropped and stretched version is only included here for the sake of completeness and as a base case against which to measure the graphical method the IMF actually chose.

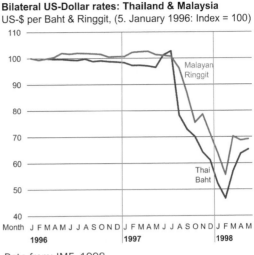

Bilateral US-Dollar rates: Thailand & Malaysia
US-$ per Baht & Ringgit, (5. January 1996: Index = 100)

Data from: IMF, 1998

The master class of axis manipulation is the logarithmic scale. Scientists tend to use it on scatter plots when there is a theoretical expectation that the relationship between the two values displayed by each data point involves squares of values. If the expectation is fulfilled by the observations, it shows up as a nice straight line, which we are good at judging by eye, connecting the dots on the plot. Non-scientists seldom use it for anything other than saving space on the vertical axis. Its condensatory capability usually leads to an apparently general graphical reduction of value differences, in much the same way as an unbroken axis suggests smaller differences than a broken axis. But in the hands of experts, a logarithmic axis can be used to achieve more differentiated effects because it selectively reduces the optical differences between large numbers more than between small, both absolutely and in percentage terms.

To return to the depiction of the US$ rates: if we not only break the axis below US$ index 40 but also scale it logarithmically, the bottoming out of the crash and the modest recovery immediately afterwards are selectively stretched much more than the minor rate variations before the crash. To put it another way, rate changes just after the IMF started to help are dramatised in comparison with rates a few months earlier, without changing a decimal point of the original data. This is what the

IMF actually did and involves the sort of graphical virtuosity that raises hairs on the nape of the neck.

As a proxy measure of PDQ we can compare the apparent size of the rate recovery of the Baht and Ringgit after their low points in February 1998 with the height of the reference value on which the index is based on 5 January 1996. Using the logarithmic as opposed to the merely squeezed-to-please scale increases the relative height of the recovery section of the curve by about 60%.

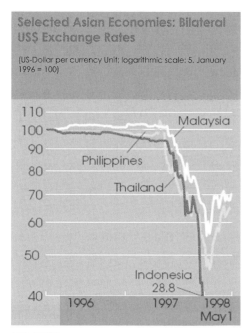

IMF, 1998

So PDQ is around 1.6:1 while the STD, gilded by the feeling of being at the cutting edge of modern econometrics, is at most zero and probably below.

Few uses of logarithmic scales are as subtly successful as this, in which the figures just happened to cry out for the method. But with careful handling they are usually good for PDQs of 1.2 to 1.4:1 with dreamlike STDs.

Enforced logarithmic scales

The logarithmic scale in the last example was blatant but still econometrically plausible. This may have been because its use seemed to have been motivated by something other than lack of space on the page. Comparison with two unsuccessful examples in which lack of space is the only apparent motive reinforces this impression.

Anyone wanting to chart energy use and its composition over time has to contend with huge value disparities. This has been true for a long time, as this example from the mid-1970s edition of the *Bartholomew/Times Atlas of the World* demonstrates. Even in full scale atlas format

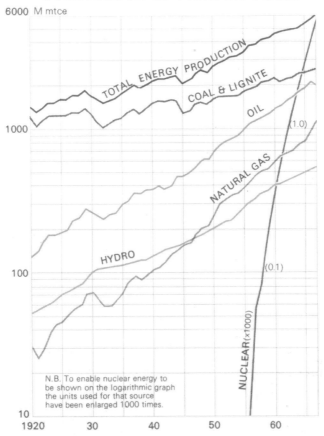

COMPARISON OF PRIMARY FUELS

6000 M mtce

Bartholomew/Times Atlas, 1975

there was no real alternative to squeezing height by using a logarithmic scale. The approximate straightness of the curves neatly makes the point that energy use has been increasing exponentially. But even then space seems to have been so tight that the scale starts at 10 rather than 1 billion tonnes of coal equivalent. This unfortunately had the side effect of pushing nuclear power, still in its infancy in 1967, clean off the bottom of the chart. The chosen solution was open, but bizarre, and involved multiplying all the figures for nuclear power by 1,000. This has the weird consequence of propelling nuclear power into rough visual parity with the sum of all other power sources, although it was in 1967 but 0.1% of the total.

This degree of visual distortion is only tolerable if the chart is designed not to be a chart at all but simply a data quarry with squiggles. On this basis, and because it is in fact possible to read off the quinquennial values for all energy types with reasonable accuracy, the chart does not entirely fail, but remains inferior to an equivalent data table.

Our last example of a logarithmic scale, obviously also motivated exclusively by considerations of space, is less successful and the reasoning behind using a non-linear scale much less obvious. The values seem to spread from just less than 0.1% to about 11%, just over 100:1, which is not an intolerably large span for a conventional chart. This is not immediately apparent in the example. The scale stretches from 0.01% of the population being in the armed services, vastly lower than the lowest, British, proportion of any of the nations shown, to 100%. The latter seems generous as it would have entailed infants being not merely in, but also under, arms. The First World War would have witnessed less the clash than the crèche of civilisations.

THE RISE AND FALL OF THE WARFARE STATE

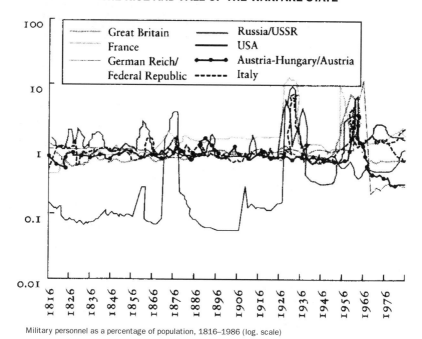

Military personnel as a percentage of population, 1816–1986 (log. scale)

Ferguson, *The Cash Nexus*, 2001

The conclusion on value scale distortion must surely be that if you use it to strengthen your message, which under favourable circumstances it achieves with bravura, you must also take care not to over egg the graphical pudding and expose the whole chart to contempt.

Bent scales

As a lunatic variation on scale condensation, you can also bend the y-axis, sometimes in several different directions at once.

Civil wars in Liberia and Sierra Leone starting in the early 1990s drove citizens from both countries into exile. By 1995 the United Nations had registered more than 3 million people as refugees. A simple x-/y-area diagram shows not only how the total number rose from

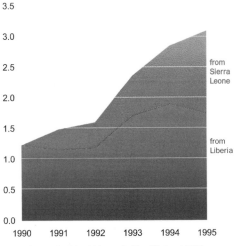

Refugees and displaced persons in West Africa 1990–1995

Data from: Smith, *Kriege & Konflikte*, 1997

1990 to 1995, but also how refugees from Sierra Leone partially 'replaced' refugees from Liberia towards the end of the same period. The figures are particularly impressive when one realises that in 2000 the total population of both countries together was less than 9 million.

It's difficult to imagine why anyone not directly involved in the tragedy (as warlord, tribal chief or corrupt bureaucrat) would want to obscure these quantitative relationships. Nonetheless, perhaps accidentally, this is precisely what the *Fischer Atlas Kriege & Konflikte* (*Atlas of War and Conflict*) managed to do.

Confusion is increased in the graphical treatment on the right by the bent arrows, which clearly suggest movement. At first sight, the chart seems to show annual flows of refugees rather than the number of those already fled, which in fact is what it's about. It's only when the thought crosses the mind that had the figures referred to movements of refugees, the problem would, in a ghastly sense, have solved itself as there would have been no one left to continue the civil war, that the truth dawns.

Unfortunately, the six different axes are so difficult to read, let alone compare, that STD is 1.0. It would be supererogatory to compute an entirely theoretical, doomed PDQ.

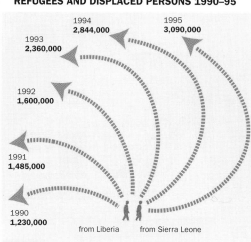

REFUGEES AND DISPLACED PERSONS 1990–95

Smith, *Kriege & Konflikte*, 1997

Multiple scales

Inconsistent scales

Whenever we are faced with numbers in more than one currency the opportunities for distortion become almost infinite, unless we translate all the currencies into one, but even then there is a menu of possible conversion methods to suit the most jaded palate. This is because we have a choice of how far to stretch the different currency axes. This beckoning world of deceit turns almost into an *embarras de richesses* if we are using prices for one commodity, say petrol, but expressed in different units of volume, in this case litres and gallons (US gallons at that). The only limitation, apparent only to the graphically innocent, seems to be that for each resulting price per unit price there ought to be a scaled relationship between the highest and the lowest value displayed. But even this can be overcome by methods we have already discussed, most obviously breaking the axis somewhere between the lowest value and zero.

This chart pair suggests strongly that the price of petrol rose proportionately by similar amounts in the USA and the UK between January 2002 and early summer 2004. The slope of the curves looks roughly the same. However the figures reveal an increase of around 63% in the USA against a mere 15% in the UK. The explanation of course is

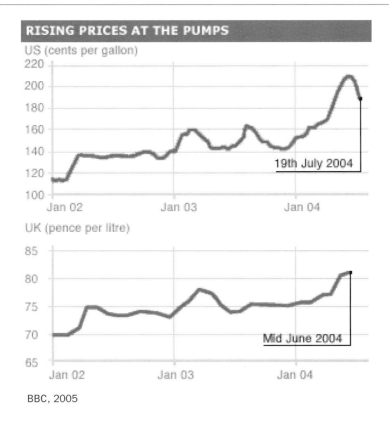

RISING PRICES AT THE PUMPS

US (cents per gallon)

19th July 2004

UK (pence per litre)

Mid June 2004

BBC, 2005

that petrol duty is so high in the UK that it takes more than a doubling or trebling of the price of the raw material to make much of an impression of the price at the pump.

The example is of more than academic interest as it makes a major contribution to international misunderstanding between the two shores of the Atlantic. The idea of going to war to secure sources of crude oil and thus to reverse a fuel price increase of 15% (as occurred in the UK between January 2002 and June 2004) would be considered as bizarre in the USA as in Europe. But for 63%, and counting, who knows?

Multiple but consistent scales

Whoever uses two rows of figures with two different scales on one chart must accept that readers and viewers are liable to assume some sort of causal relationship between the two rows. This is encouraging for the potential graphic deceiver. Contemplation of the number of charts, in which axis fraud can get two such curves to run roughly parallel to one another for long periods can be a deeply revelatory experience. The feeling of causality is heightened when the change in one series can be made to look roughly the same size as the change in the other

In 2005, as in 1991, there were roughly 40 million economically active (employed plus those actively seeking employment, i.e. the unemployed) people in Germany. As the total population grew slowly over the same period the proportion of economically active people declined slightly. According to *Der Spiegel* magazine, however, the big change between 1991 and 2005 was in unemployment, which rose by 2.6 million, the number in employment falling obviously by the same amount. There was also a strong increase in employment in service industries and major declines in manufacturing (3.7 million) and, to a lesser extent, in agriculture. To put it another way, rises in service industry employment failed to compensate fully for declines in industry and agriculture.

But what if you want to write an abrasive article about the things governments do that make Germany less internationally competitive in manufacturing? Wouldn't it be nice to make the rise in unemployment in general (2.6 million) look as much as possible like the fall in employment in manufacturing in particular (3.7 million)? Simply comparing the 2.6 and the 3.7 reduces message purity and polemical impact because it suggests that the government was getting something right, namely the rise of about 1.1 million in employment in the service sector. So how do you get 2.6 to look roughly like the 40% greater 3.7?

The answer of course is to use different scales for decline in manufacturing employment and rise in unemployment, in spite of the fact that both the time periods (1991–2005) and the value units (people) involved are the same. Throw in some mild misalignment of the resulting twin charts but connect up their supposedly zero base lines and increase optical confusion by replacing grid lines with an irrelevant wire model globe and you're home and dry: decline in manufacturing caused unemployment; end of story.

As twin axis distortion goes, this is a mild, though competently executed, example with a PDQ of around 1.4:1. The STD is difficult to judge. At first glance it's not too bad, though the connected x-axis is a gratuitous provocation, but the discount climbs steeply as we try to draw our own conclusions from the chart, probably ending up on second

Employment in manufacturing and total unemployment (millions)

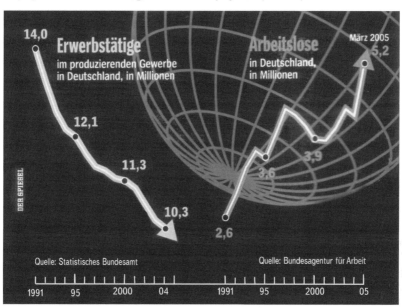

Der Spiegel, 2005

glance at around 0.8. As a general rule using multiple axes, especially for one and the same value unit, leads to such high STDs that it all seems scarcely worthwhile.

It's fitting to end this chapter with a twin axis chart from a brilliant if contentious book that shaped international geopolitical discourse for a decade: *The Rise and Fall of the Great Powers* by Paul Kennedy.

Few devices have been omitted to make the graphic as impenetrable as possible and even Professor Kennedy is moved to admit that it 'may be too schematised a presentation to most readers.' The legend wastes so much space below the graphs that it reflexively justifies itself, the meaning of the measure (r^2) – presumably of the quality of fit of the trend lines – is unexplained and the connection between Kennedy's conclusion that 'the 'relative power' of Russia was just rising from its low point after 1894 whereas Germany's was 'close to its peak' is masterfully obscured by placing the two graphs side by side rather than one above the other. The German dots seem more fitted to some sort of cata-strophic chaos theory than the relatively smooth trend line suggests. The coup de grace is delivered by the different labelling of the two value

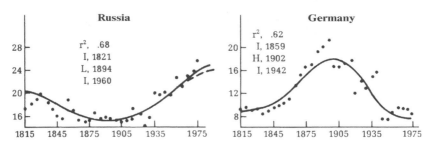

The Relative Power of Russia and Germany

Key:
L = year of low point
H = year of high point
 I = year of inflection point

Kennedy, *The Rise and Fall of the Great Powers*, 1988

axes, which raises doubts in the reader's mind about the comparability of the two curves.

Even if the data we have chosen to put into the chart are impeccably neutral, the available techniques for disguising graphical length, for distracting from real values with icons and the use of irregular axes mean that the final impression made by the graphic's value axis is entirely in the hands of its author.

Thus encouraged, it is time to turn to the category or time axis of the chart and its manifold techniques of distortion.

5 DISTORTING CATEGORIES AND TIME

Data manipulation

A s with the value axis we can also divide manipulation of the other (category or time) axis into techniques that concentrate on the data before it gets as far as the chart (data manipulation) and those that directly bend the optical impression created by the chart (graphical manipulation). Both have a major contribution to make to adjusting reality and are often successfully practised in parallel. Sticking to the sequence in which most charts are actually created, let's start with data manipulation, in particular with all the wonderful ways they're using definitions nowadays.

Manipulating category definition

The way categories are defined can make all the difference to the resulting graphic. At the heart of most of the deception techniques that exploit definition is usually a readiness to allow readers and viewers to revert to their own definitions. As in most cases we can make a reasonably accurate forecast of what these definitions will be, it only remains not to obtrude our own definitions on the reading public and so run the risk of achieving unwanted clarity.

Omit crucial part of category definition

Suppose we wanted to show that the government was 'still trying to educate on the cheap'. The most obvious statistic to show would be expenditure per pupil/student compared to other industrialised countries. But even if you use purchasing power parities rather than £s or €s (to correct for different costs of living) this direct approach raises a whole lot of comparison problems. It's much less contentious to compare the proportion of national income that governments spend on education.

Defining the data this way however gives only rather pallid support to our thesis. According to the Organisation For Economic Cooperation and Development (OECD), in this as in most other areas of public policy in rich countries the acknowledged international data source, the UK goverrnment spends a slightly smaller proportion of GDP on education than the average OECD country. The USA, for example, spends very slightly more. This may get them

Public expenditure on education in % of GDP 2001

Ireland	4.1%
Germany	4.3%
Netherlands	4.5%
Australia	4.5%
UK	4.7%
Korea	4.7%
OECD average	5.0%
USA	5.1%
France	5.5%
Belgium	6.0%
Sweden	6.3%
Denmark	6.8%

Data from OECD, 2004

excited in Whitehall and various staff rooms and Senior Common Rooms throughout the land, but the difference between 4.7% and 5.0% is not the sort of stuff that gets Jacobin tricoteuses crowding round the foot of the guillotine.

In case an alert reader is surprised by the low American figure, note that this is public, not total, expenditure on education. The total expenditure figures are given in the table here. The Korean and USA figures are increased enormously by including private expenditure on education; the UK figures are also slightly improved, being now but 0.1% points below OECD average. But our readers may feel that this second table

Total expenditure on education in % of GDP 2001

Ireland	4.5%
Netherlands	4.9%
Germany	5.3%
UK	5.5%
OECD average	5.6%
Australia	6.0%
France	6.0%
Belgium	6.4%
Sweden	6.5%
Denmark	7.1%
USA	7.3%
Korea	8.2%

OECD, 2004

is slightly beside the point, if what we are trying to demonstrate is ill-judged or inappropriate government parsimony.

So we must try an indirect approach. Perhaps we can keep the figures from the top table but show at the same time that in the case of the UK the money is being spread so thinly, divided among so many bog-standard overcrowded classrooms, that universal illiteracy may be less than a generation away.

A few pages and data tables further on at the OECD, *The Times* struck pay dirt in the form of 'education expectancy' which appears to

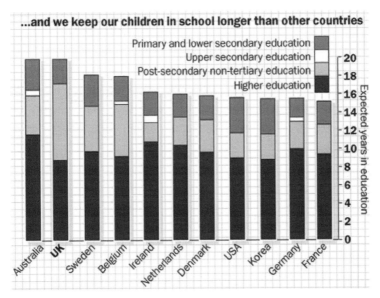

...and we keep our children in school longer than other countries

The Times, 2005

show, as *The Times* puts it, that 'we keep our children in school longer than other countries'. Ignore the irritating typographical error (the order of labelling in the legend has been inverted), which it would be childishly adversarial even to mention. So pausing only to remark that illiteracy may already be more widespread than *The Times* itself appears capable of correcting, we come to the startling suggestion that the average UK child spends almost nine years in upper secondary school (roughly speaking: secondary school beyond the end of compulsory full-time schooling, shown by the light grey portion of the bars above), markedly longer than any other country shown.

This is the core of the distortion because nothing on the chart points out that more than half of this upper-secondary-school expectancy in the UK is part-time. As the chart on page 116 shows, only Australia shares this characteristic with the UK. Indeed over a quarter of *total* school expectancy in the UK is part-time and, to put it kindly, of unconvincing quality. But, ignoring quality factors, *The Times* achieves here a PDQ of around 2:1 measured in terms of the exaggeration of the length of UK schooling compared to the OECD average. In its chart on total expectancy the UK comes equal top with Australia. In reality, on full-time education expectancy the UK comes bottom.

Expected years of schooling (full- and part-time), 2002

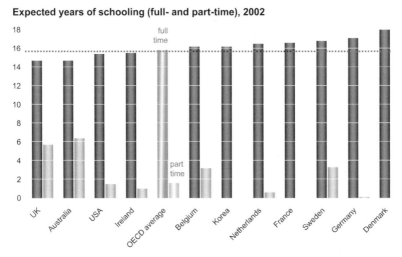

Data from: *The Times*, 2005

The Times then completes the package by adding two further charts. The top chart shows that young people will remain a greater proportion of the population than old people until 2020, which, while true enough, is distracting as the population share of young people (the main target of education expenditure) will continue to decline until 2050 along with the costs of educating them. The bottom chart, of similar layout to the expectancy chart and hence inviting comparison more directly, shows total national expenditure on education as a % of GDP. It too contains figures not in the cited source, but simply increasing the height of the USA and Korea bars and shifting them to the far left, moving the Netherlands bar to second from right and ignoring the scale would restore a modicum of verisimilitude. By omitting a crucial part of a category definition, in this case 'expected years in education' includes an unusually large proportion of part-time education, and by juxtaposing it with the flawed portrayal of actual education spending, *The Times* indeed does suggest that too little money is being spread too thin.

Though a bit over-elaborate, this graphical saga overcomes consider-able data restrictions to reach a triumphant conclusion. The typos provide a handy diversion, but the core of its success is to withhold a crucial part of the definition of what it means by 'Upper secondary education' and fail to explain why this omission makes such a difference to the impression we get from the chart as a whole.

STILL TRYING TO EDUCATE ON THE CHEAP

Young people in Britain will continue to outnumber the elderly until 2020...

Elderly and young as a proportion of the working population

Proportion of old people relative to the working population

Proportion of young people relative to the working population

100%
90
80
70
60
50
40
30
20
10
0

1960 1970 1980 1990 2000 2010 2020 2030 2040 2050

Source: GAD 2002-based principal population projection, UK. ONS Population estimates unit, UK

...and we keep our children in school longer than other countries

Primary and lower secondary education
Upper secondary education
Post-secondary non-tertiary education
Higher education

20
18
16
14
12
10
8
6
4
2
0

Expected years in education

Australia UK Sweden Belgium Ireland Netherlands Denmark USA Korea Germany France

Yet Britain's spending on education remains unusually low

9%
8
7
6
5
4
3
2
1
0

Percentage of GDP

Denmark Sweden Belgium France USA Canada Australia Netherlands Korea UK Germany Ireland

The Times, 2005

Obscure the context and wording behind polling results

Opinion polling involves not only composing questions and multiple choice answers for the questionnaire but also a choice of the categories to use when we report the results. This is particularly fertile ground for us because it would be unreasonable to expect a newspaper to find space for all the nuances and contextual details of one, let alone several, slightly differing polls. Yet, as we shall see, barely discernable differences in wording appear to have deeply satisfying effects on the answers we get. The chance of anyone having either the time or the inclination to visit the websites of the polling organisations to find out what was really going on on a thousand doorsteps three weeks before publication is so remote as to be negligible.

In May 2005 *The Times* published the results of eight opinion polls taken over the previous year on the subject of the proposed European constitution. There is no reason to think that the results were rigged, in the sense of inadequate sample sizes or biased sampling techniques. The polls were conducted by well-known and reputable polling organisations, even if for newspapers and political parties with yet better-known, and in some cases extreme, opinions on the issue. The results show great volatility of public opinion on the subject during this period, and it is highly probable that some of this volatility was genuine. But a glance at differences among the polls suggest that some was an artefact of the wording of the questions that were asked and of the context in which they were asked. In one case, an apparently inconsequential change of wording, all other factors remaining unchanged, produced such utterly different results that it should go down in legend and song as the graphical answer to awkward poll results.

Let's start with the context in which the crucial question is put, a subject explored in hilarious detail by Sir Humphrey Appleby (in the context of a lunatic plan to gain opinion poll support for the return of National Service) in the comedy series 'Yes Minister' half a generation ago.

By a clear margin, the most negative net reaction to the proposed EU constitution was produced by poll 1, conducted between 15 and 17 April 2004 by YouGov for the Sun newspaper. The sample of 2,462 electors was unusually large, the poll was carried out online (admittedly itself a possible source of error, though YouGov chose the sample) and the

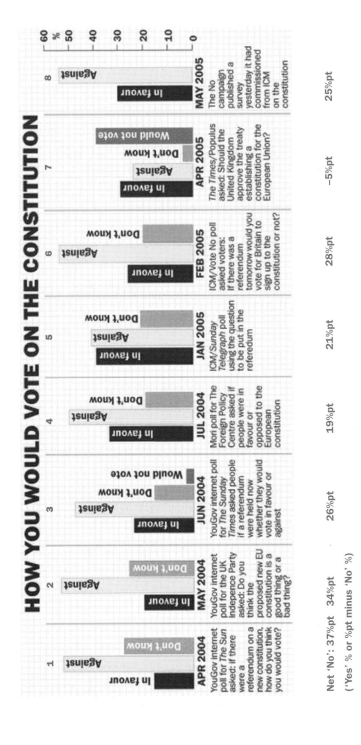

HOW YOU WOULD VOTE ON THE CONSTITUTION

APR 2004
YouGov internet poll for The Sun asked: if there were a referendum on a new constitution, how do you think you would vote?

MAY 2004
YouGov internet poll for the UK Indepence Party asked: Do you think the proposed new EU constitution is a good thing or a bad thing?

JUN 2004
YouGov internet poll for The Sunday Times asked people if a referendum were held now whether they would vote in favour or against

JUL 2004
Mori poll for The Foreign Policy Centre asked if people were in favour or opposed to the European constitution

JAN 2005
ICM/Sunday Telegraph poll using the question to be put in the referedum

FEB 2005
ICM/Vote No poll asked voters: If there was a referendum tomorrow would you vote for Britain to sign up to the constitution or not?

APR 2005
The Times/Populus asked: Should the United Kingdom approve the treaty establishing a constitution for the European Union?

MAY 2005
The No campaign published a survey yesterday it had commissioned from ICM on the constitution

Net 'No': 37%pt	34%pt	26%pt	19%pt	21%pt	28%pt	–5%pt	25%pt

('Yes' % or %pt minus 'No' %)

The Times, 2005

question neutral: 'If there were a referendum on a new constitution, how do you think you would vote?' Respondents could choose between 'In favour': chosen by 16%; 'Against': 53%; 'Don't know': 28% and 'Would not vote': 4%, giving a net 'No' result of 53% – 16% = 37% points.

The Times summary however fails to reveal that this was the last of five questions asked. The stage is set by the first, apparently irrelevant, question: 'Thinking now about Britain's relations with the rest of the European Union, do you believe the following statement is true or not true: 'The European Union wants all member states, including the United Kingdom, to sign up to a binding constitution'.'

Only a continental nitpicker, noting that before a referendum or some other ratification process in all member states no one could know what the EU really wanted, could possibly disagree with the statement. Only 9% did.

'The European Union wants ...', 'sign up to ...' and 'binding ...'; you can already see Napoleon's invasion barges massing at Boulogne and hear the opening chords of Vera Lynn and the White Cliffs of Dover. For those who still didn't get it, the third question turned up the volume: 'Do you think the following statement is true or untrue: 'Britain will still be able to keep control of its taxes, defence, criminal justice and foreign policy if a new European constitution does come into force'' while the fourth points to the only escape route from this catastrophe: 'If Europe's leaders do agree a new constitution for the European Union, who should decide whether Britain signs it? The decision should be made by Parliament/by the people in a referendum/don't know.' Then, and only then, is the crucial question put.

It's moving to see life imitating art and full marks to *The Times*, and to its stablemate the *Sun*, for not spoiling the joke by revealing the context of the question.

After this triumph, however, support for the 'No campaign' seems to have ebbed, the net 'No' declining successively to 34, 26, 19 and 21% percentage points in various polls by January 2005. But suddenly in Feburary 2005 (poll 6), some clarion call seems to have awakened the slumbering British public to its deadly peril. The question put to 506 adults aged 18+ by telephone between 2 and 3 Feburary 2005 by ICM on behalf of 'Vote No' was: 'If there was a referendum tomorrow, would you vote for Britain to sign up to the EU constitution or not?'. Note the abandonment of the pedantic 'If there were …' in favour of the more demotic 'If there was …' Possible answers were 'Vote to join': chosen by 26%; 'Vote not to join': 54% and 'Don't know': 20%, restoring the net 'No' to a respectable, even unassailable, 28% points.

Wisely, both the 'Vote No' campaign and *The Times* chose not to publicise the results of another question asked of 522 respondents in a parallel poll also conducted by ICM for the 'Vote No' campaign under the same conditions on the same date. The question was apparently only slightly different: 'Should the United Kingdom approve the treaty establishing a constitution for the European Union?' allowing the answers 'Yes': 39%; 'No': 39% and 'Don't know': 22%, producing a net 'No' of precisely zero.

Given that the answer doesn't lie with the 'Don't knows' who polled almost the same in both cases, what can explain the difference? There seem to be three possibilities: random variation (which is statistically highly unlikely), some curious potency of the phrase 'sign up to' compared to 'approve' and the fact that the first formulation enquired after a voting intention (though without allowing a 'Won't vote' response) and the second after a mere opinion.

Whatever the cause, by February 2005 *The Times* and other luminaries of the 'No campaign' could have been forgiven for thinking that the results of a referendum on the proposed constitution were by no means a foregone conclusion. Poll 7 by Populus for *The Times* itself on 22–23 April 2005, closing some of the interpretational loopholes in the February poll, would have heightened their unease. The question this time was longer and the sample, at 711, larger: 'Next year a referendum is to be held on the proposed EU constitution. The question on the ballot paper will be 'Should the United Kingdom approve the treaty establishing a constitution for the European Union?' How would you vote if this referendum were taking place now – or would you probably not vote at all?' Note that the official question on the ballot paper was going to use 'approve' rather than the possibly more inflammatory 'sign up to'. Is there a hint of desperate invitation in that 'probably not vote at all'?

In the event, the results could scarcely have been more disturbing: 29% responded with 'Yes'; 24% with 'No'; 39% 'Would not vote' and 8%

'Don't know' giving a net 'Yes' of 5% points. The eighth poll cited above, by ICM for the 'No campaign' in May 2005 seems more than to reverse these findings, but the details of the question and its context are no longer available either from the now moribund 'No campaign' or from ICM.

The whole question was rendered void a few days later, as so often in modern British foreign policy, by decisions taken by other nations, in this case the French and Dutch. Both are said to have voted 'No' – perhaps in some confusion of mind – to exclude Polish plumbers, as opposed to the home-grown variety, from fondling the pipes in their bathrooms. True or not, the proposed Constitution, which would not in fact have conferred the slighest additional privilege on Slav handymen, was now dead.

The lessons for those who practise to deceive with opinion polls could not however be clearer: get the context and the wording right and if you can't be sure what's right, use two and pick the right one when the results come in.

Of course, if you are in the happy position of being able to define your own sample, a lot of this pussyfooting can be abandoned. Take for example a recent pamphlet by the Boston Consulting Group, which like many others if the same ilk is never shy of using its American origins in

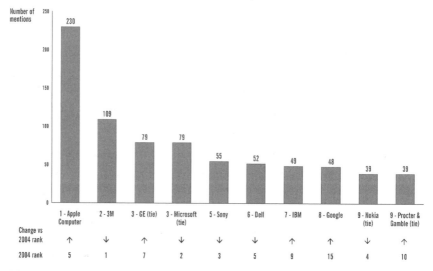

THE 'MOST INNOVATIVE' COMPANIES

BCG, *Innovation*, 2005

subliminal sales arguments. In this case the tactic seems to have been to alarm European executives at their own lack of innovative zeal compared to Americans. From this of course it is but a short step to hire specialist consultants from the more creative culture to put it right. In this case some 940 senior executives were polled world wide – the methodology page in the pamphlet lists an impressive 28 different countries for the respondents plus 'other/unspecified' – on their choice of the world's most innovative companies. Out of the ten judged most innovative, eight were American, one European and one Japanese. Assuming, of course, that all 940 were saintly men and women, unaffected by questions of national pride, let alone commercial advantage, it would probably still remain true that local champions would come more readily to mind than those from overseas. But BCG's list of nations of respondents (it is not entirely clear whether residence, origins or passport is meant, but the implication seems to be nationality) most unwisely reveals 43% under 'U.S.' against 21% under the various states of the EU. However, the methodological explanation is well separated from the chart and most time-stressed executive readers can be relied upon to miss the nuances of the sample.

The key to this technique is to avoid the temptation to obscure your polling methods entirely, but to ensure that an apparent flash of honesty reveals as little as possible as unobtrusively as possible. Nice one.

Change definition of 100%

Charts are often used to show parts of a whole. In fact, there's not much else you can legitimately do with a pie chart, though the illegitimate uses, attempting for example to compare the segements of more than one pie, can be quite fun too, as we have seen. This does not mean that single pies offer no opportunities for deceit.

One of the crucial, and often satisfyingly unanswered, questions about pie charts is: what's the whole 100% of which we are being shown various shares? A slightly scatty, but somehow sympathetic, American outfit called the War Resisters' League (WRL) provides a topical example in which this question is explicitly at the core of the problem. According to the WRL the US Federal Budget looks like this, with almost half being taken either by current military expenditure or mopping up the results of previous military expenditure in the form of debt repayment, interest on the debt incurred or payments to ex-servicemen, who managed to escape the military with their pensions and lives intact. But as you go into the details it becomes clear that special pleading is being used on a scale to make Enron's last annual report look as regular as tide tables.

The first and most important distortion is what has been excluded from the 100%, in this case from the budget of the US federal government, which official sources portray like this. WRL excludes so-called trust funds like Medicare and Medicaid on the grounds that the funds for these health programmes for the old and poor are not financed out of general taxation but by dedicated deductions from paycheques. This also excludes most of social security. While it makes little difference to the individual exactly how the government grabs the money, the WRL graphic elegantly sidesteps the problem by using the headline 'Where your income tax money really goes'.

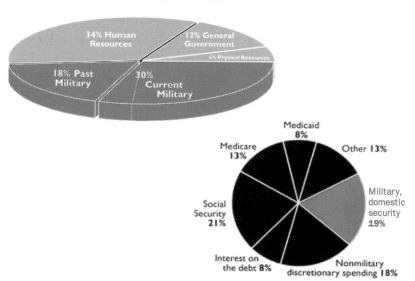

War Resisters' League, 2005

124

The second method, which sounds more reasonable to European ears, is to include in military expenditure money spent on all clearly military items, for example nuclear bombs, which in an inspired and time-hallowed bit of Newspeak are paid for by the US Department of Energy.

Third, and perhaps most controversially, WRL declares that 80% of interest on the federal debt, easily the fastest growing item of the entire budget, should also be classified as military spending. Federal debt is the simple result of the government spending more in total than it takes in as tax revenue. As Mr Micawber put it: 'Annual income twenty pounds, annual expenditure nineteen nineteen six, result happiness. Annual income twenty pounds, annual expenditure twenty pounds ought and six, result misery.' Attributing the 'ought and six' to any one particular group of programmes, for example defence, is even sillier, and quite a lot more controversial in economic and accounting theory, than attributing parts of your mortgage interest payments to particular tiles in the kitchen floor. After all, even the WRL is 'only' suggesting that half of federal expenditure is military, so charging the military with 80% of the interest on the national debt looks rather greedy.

The verdict is mixed on WRL's accounting adjustments and this book has no intention of inviting terminal prejudice by attempting to adjudicate a dispute in a far off land of which we know little. What remains for our purposes is the neat little trick of using what seems at first to be a mere rhetorical device – 'your income tax' – in an anyway OTT headline to redefine the meaning of the entire graphic.

The PDQ of this adjustment is around 2:1, the STD is probably low (don't forget that paying income tax in the USA involves all the dread and awe of actually filling out and signing an annual cheque to the government and so looms large in the American mind compared to social security and so on, which are deducted at source) and so it's one to treasure.

'% of what' on same graph

A slightly more graphically sophisticated version of the same 'spot-the-100%' technique depends on using percentages of percentages to undermine or obscure meaning. In this example the original data difficulty has been compounded by an elegantly distorting graphical treatment. The height of the Olympic rings – hey, neat idea – is proportional

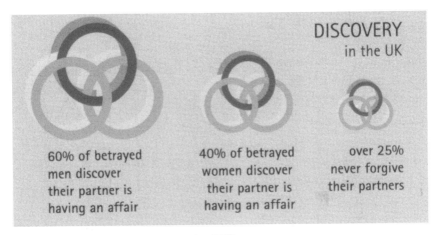

DISCOVERY
in the UK

60% of betrayed
men discover
their partner is
having an affair

40% of betrayed
women discover
their partner is
having an affair

over 25%
never forgive
their partners

Mackay, *Atlas of Human Sexual Behavior*, 2000

to the percentage figures below, ignoring the fact that the 25% for those who never forgive their partners is presumably related to the 60% and 40% of those who discover that their partner is having an affair. If so, it should therefore be around 50% smaller than it is. On the other hand, when you come to think about it, it's not impossible that it refers to that quarter of all married people who eventually forgive their partners nothing, above all having married them in the first place. However, in that case the rings on the right are distinctly undersized. On top of all these problems there is the usual graphical ambiguity with icons, about whether we should be looking at the height or the area of the graphical elements themselves and in this case whether we really need any graphical treatment at all.

All in all, a successful little chart that manages to submerge its already modest information content in a wave of reader irritation.

Use overlapping categories

A lot of the manipulation of category definitions involves in some way the violation of the MECE rule, that categories should be Mutually Exclusive and Completely Exhaustive. For example, not the least effect of confusing category definitions is to make it difficult to check if the categories and data are completely exhaustive or not.

But the mother of all category fraud is using categories that are not mutually exclusive. This deliciously gross example from the *Penguin Atlas of Human Sexual Behavior* sets a standard that few are destined to achieve. It would be nice to think that it is the result of some *jeu d'esprit*, some wager involving an attempt to produce the worst graphic that the mind of man can conjure up, but the internal evidence points more boringly to simple incompetence on the part of a poorly educated North American. An estimate published in 1994 by Penguin itself came up with a figure of around 20% for white

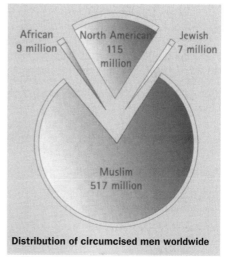

Distribution of circumcised men worldwide

Mackay, *Atlas of Human Sexual Behavior,* 2000

UK males, which would point to at least 4 million non-Jewish, non-Muslim circumcised males in Britain, let alone the rest of Europe. There are several significant Christian churches outside Europe which urge circumcision, not to mention significant non-Muslim religious and ethnic groups in Asia and Oceania. It is not surprising therefore that the total in the chart above (648 million) falls distinctly below the 'somewhat over 1 billion' that seems to be the usual worldwide estimate.

From our point of view, violating the mutual exclusivity constraint can be a useful way of distracting from, even making impossible, rational consideration of the figures so presented.

Unusual category definitions: the Disney illusion

A lot of recent economic journalism depends upon instilling a healthy panic in the Western reader about the approaching disappearance of Western economies from the list of economic giants. Various techniques are used.

One of the most popular is the overextended trend line, much beloved of the Club of Rome and most subsequent doomsayers, which happily extends the economic trends of the last ten years for the next 50 or more. As the Duke of Wellington replied, when hailed in the street as Mr Smith, 'if you can believe that you can believe anything'. At the time of writing, this method would predict Ireland as the biggest European economy by the middle of this century. Another, related, technique is 'never mind the price, feel the quantity', which assumes the supension of price effects on demand until after the author's favourite catastrophic shortage – energy, food, water etc. – has come to pass. Most such techniques rely heavily on a shortness of their audiences' memory bordering on clinical amnesia. A previous version of the present teach-your-children-to-count-in-rupees school of economic forecasting in vogue in the 1980s predicted, for example, that Japan would overtake the EU and the USA in absolute economic size about now.

No one disputes that the Chinese and Indian economies are growing fast and will probably continue to do so for many years. But, if panic is your goal, it helps to reduce the apparent size of their Western competitors and here category definition can make a contribution. In this example from Germany's leading business weekly, the shrinking European share of the world economy, the top segment of the bars will have fallen to less than 10% by 2025, compared to 35% in 1913. The American share in 2001 is shown as being almost twice as large as the European, though in fact on OECD

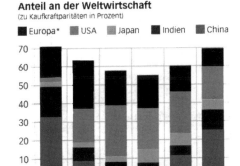

Shares of the World Economy
(in %, based on purchasing power parities)

Anteil an der Weltwirtschaft
(zu Kaufkraftparitäten in Prozent)

■ Europa* ■ USA ■ Japan ■ Indien ■ China

*Deutschland, Frankreich, Großbritannien und Italien:
** Prognose; Quelle: Virmani

WirtschaftsWoche, 2005

figures the gap was only about 10%. At this stage the reader starts searching the chart for explanations and, sure enough, the first footnote announces the creation of a Bonzai-Europe consisting only of Germany, France, the UK and Italy. The PDQ in this category definition is a handy 1.6:1 today and grows rapidly in the future as the states omitted in this EU definition are also easily the fastest growing. STD is low as the footnote is unobtrusive and, even if read, does contain the names of the four biggest member states.

The STD of this chart is probably particularly low in Britain where the Disney illusion is most common. The Disney illusion is named in honour of Walter Elias Disney, the creator of the cartoon version of Snow White. The key point is the number of dwarves. In 1956 it was six, but rose first to nine, then 12, 15 and since 2004 has reached 25 upon the accession of ten Eastern European and Mediterranean states to the EU. Snow White, of course, is the USA, which presumably makes George W. Bush either the wicked step-mother or the poisoned apple. The illusion works by disaggregating European figures for anything from GDP to trade and even aircraft carriers into numbers for each individual state but presenting the USA as a monolithic whole. It tends to inflate the importance of the USA on the world stage and is an almost exclusively Anglo-Saxon phenomenon.

Misleading definition

Although we tend to believe less in the march of human progress these days, it's always nice to see some figures that suggest that it still goes tramping on. This chart shows the spread of the female franchise from its inception in that hotbed of women's lib, New Zealand, to most of the rest of the world in the

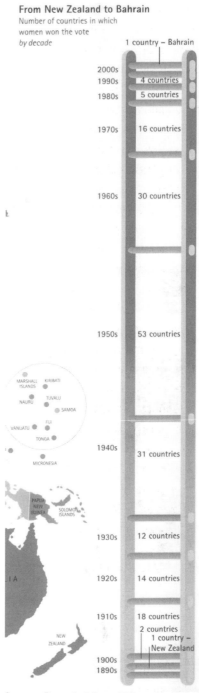

From New Zealand to Bahrain
Number of countries in which women won the vote by decade

1 country – Bahrain

2000s	
1990s	4 countries
1980s	5 countries
1970s	16 countries
1960s	30 countries
1950s	53 countries
1940s	31 countries
1930s	12 countries
1920s	14 countries
1910s	18 countries
	2 countries
	1 country – New Zealand
1900s	
1890s	

Seager, *Penguin Atlas of Women in the World*, 2003

following hundred years. It's a stirring tale and one of the few that the campaigners and the general publics involved can look back on with a modest pride in its uplifting message

But can they? Careful inspection of the figures reveals that since about 1970 the increase in the number of UN member states with female franchise has been almost exclusively provided by the rise in the total number of members of the UN (i.e. in increasing numbers of newly independent ex-colonies), most of which had already had female franchise in the limited democratic institutions permitted by their then colonial masters. A group of long-independent states without female franchise has remained substantially unchanged for decades. To put it another way, while democracy in general and especially independence for former colonies has swelled the ranks of UN member states with universal franchise, the female half has experienced almost no relative improvement in enfranchisement at all. In this extended and redrawn version the gap between the top of the black bars (independent states) and the red bars (independent states with female franchise) barely changes after 1970. Add to this the fact that in many nominally enfranchising member states women do not in practice get elected to legislatures and may often not even be allowed to vote by social, religious and family pressures, and the belief in continuing progress begins to crumble.

Number of independent states with female franchise, 1900–2005

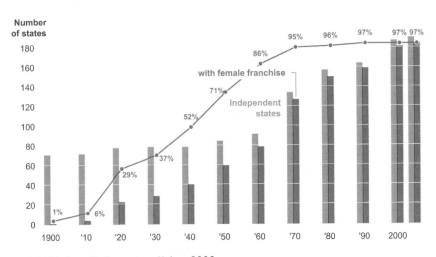

UNO 2006, Inter-Parliamentary Union, 2006

This neat little manipulation could also have been filed under the more general technique of the exclusion of data perspective, in this case of the changing number of independent states as such. The technique usually employs percentage or index figures to sever the link to absolute reality, but in this example lonely absolute figures, numbers of newly independent states with female franchise, have been quietly detached from their context and produce a significantly distorted impression.

Manipulating category selection

Even if the categories are defined clearly and the definitions are not too much at odds with everyday usage, we still have the opportunity to make a careful selection of the data categories we choose to display in the chart. This is obviously true in the case of a time series, where the provision of historical perspective must be handled with care, but it also applies, for example, to selecting representative member states for various theories about the EU or the choice of suitable price indices with which to compare the prices of Britain's leading supermarket chains.

Select historical perspective

Providing historical perspective is all very well, but it takes up more space on the page and always raises the awkward question: how much? For our more selective approach the only answer can be: as much as proves our point but not so little that the plausibility of the entire chart is terminally degraded.

Take for example the recent panic sweeping through several industrialised countries – not least the UK and Germany – about the ageing of the population and the corresponding angst about how many of working age will be left to support us in our old age. Opinions differ about the scale of the coming disaster and depend crucially on which base year we choose from the past as offering an acceptable or still viable age structure.

For Germany a judicious choice of historical perspective is particularly important. Going all the way back to 1880 – almost to the foundation of the modern German state in 1870 – threatens to put the age structure in 2050 so strongly in perspective, that neutral observers might wonder what all the fuss is about. Even going back only to 1973, the height of Germany's post-war economic miracle, when the potentially productive

% of German population aged 15–65

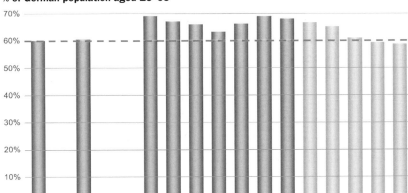

Statistisches Bundesamt 1977 and following

share of the population was only 63% (as opposed to a slightly lower 59% forecast for 2050) robs the latter figure of most of its terror. The way round this difficulty is two-fold.

First, it's important to use the dependency quota (the number of those too young, <15, or too old, >65, to work per potentially productive 15-65-year-old) rather than the 15–65 year old share of the population. This is simply because in the ratio figure, non-productives per productive, the numerator and the divisor are set to change in opposite directions. In the share figure on the other hand, both fall after about now, so the percentage change is less impressive.

Dependency quota 1975–2050
Dependents per economically active person

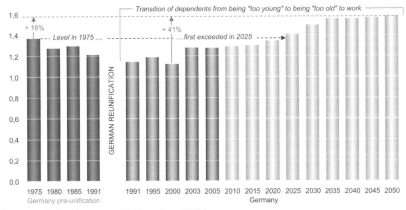

Strange, *Keine Angst vor Methusalem,* 2006

But the deletion of all years before 1975 is also useful and the remaining figures from before 1990 (the ideal starting point since it marks the highest share of working age Germans since 1945) can be pushed into the background on the excuse that they only refer to the former Federal Republic without East Germany. The final result is a bit weaselly, but manages to focus attention on the most appropriate comparisons for the message of the chart. Based on the different change rates between base years of 1975 and 1990 and 2050, the PDQ is over 2:1 and the STD approaches 0 as the historical perspective seems on the face of it quite generously chosen.

Omit relevant but contradictory categories

How fast the economies of the individual EU states have grown in the last decade is largely a function of how poor they were at its start. Most poor states have grown fast; rich states (with the small, but notable, exceptions of Ireland and Luxemburg) have grown more slowly. So far, so uncontroversial, especially as this sort of catching up by poor acceding states has been one of the main objectives of the EU since 1990. Luckily however, being poor also means that such states are not yet in a position to afford extensive and expensive social safety nets for their citizens and so can be classified as 'liberal economies' as opposed to the 'social model economies' of the richer states. For politicians in

GDP/head in 1997 (index EU-25 = 100) and GDP growth 1996–2005 (%)

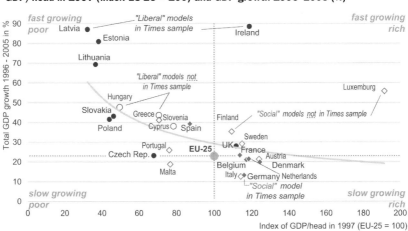

Eurostat, 2006

133

(until recently) poor states, such as the UK (which only attained the average income per head of the rest of the then EU in 1997) or Ireland, who are debating the extent of the social safety nets appropriate to their economies, this double correlation between recent poverty, growth rates and social provision is too good an opportunity to be missed and the way it has been exploited here should be a graphical inspiration to us all.

Anyone who produces a chart about economic policy in the EU but only shows the figures for 12 of the 25 member states is probably up to some sort of distortion based on a carefully chosen sample. Ignoring both the weakness that one year's growth rate is a rather narrow basis for evaluating such fundamental economic principles and the bold distortion inherent in giving the Latvian and German economies equal optical weight, this example from *The Times* makes brilliantly selective use of the double correlation. Of the seven 'liberal economies' cited, five enjoyed less than half the average EU standard of living in 1997 and have yet to achieve 60%. Of the six rich states with above average growth, four were in the 'social model economy' camp, but only the least dynamic – France – has been included in *The Times* sample. The argument that the other three were too small to be of interest might just excuse the omission of Luxemburg but still lacks conviction, not

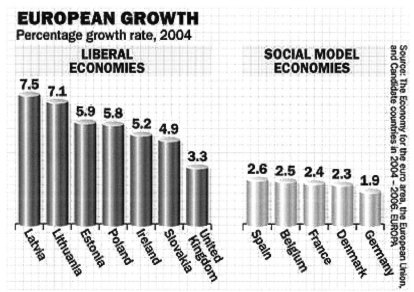

The Times, 2005

least because the omitted Swedish economy was larger than that of Latvia, Lithuania, Estonia and Slovakia (all included in the chart) combined. Even the choice of a total of 12 sample states is inspired. For older readers it echoes the actual size of the entire European Community (as it then was) some 20 years ago and helps to suggest that omissions are negligible.

This is neither an economics textbook nor a campaigning tract and so takes no side in this complex debate. In any case, as in most disputes about Britain's European Problem, neither side seems anxious to expose its foetid arguments to the cleansing light of mere reason or fact. It could be that *liberal* economies really do grow faster or, on the other hand, that they only seem to, because most poor states grow faster and couldn't yet afford to be 'social model' economies even if they wanted. So our admiration is strictly reserved for this particularly fine and unfoxed edition of the category choice technique being used to exclude awkwardly contradictory evidence.

Select the price index

The discussion about the costs and benefits of the rise and rise of supermarket chains such as Tesco continues. We hear a lot about how smaller food retailers are being driven to the wall by allegedly predatory pricing, how farmers are being driven into penury by ruthless purchasing policies and how England's green and pleasant land is being defiled by

Price inflation by main area 2000–2006
Retail price index (rebased from January 1987 to January 2000 = 100)

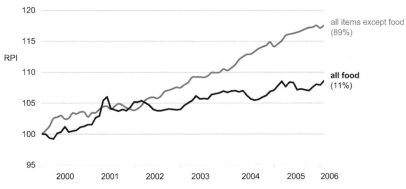

Data from: National Statistics, 2006

car parks full of discarded plastic wrapping and maimed trolleys. The short retort is that all this wouldn't be happening if customers didn't like the result, though admittedly the same could be said of the spread of crack cocaine in our inner cities. At a slightly more sophisticated level, it is claimed that the intense competition among a small number of national food retail chains has forced our food distribution systems to become so efficient that both customers and supermarket shareholders have been generously rewarded: the first by lower food prices, the second by higher dividends.

Unfortunately the statistical evidence for this supposedly textbook example of the price benefits of competition to consumers is thin, although it starts off quite promisingly. Between 2000 and 2006 the average retail price of all food rose by about 8%, which sounds pretty reasonable, especially if we compare this modest rise with the rise in all other items in the retail price index (including services, housing and tax) of almost 18% over the same period. As the supermarket chains have a much higher share of the food than the non-food market, it could be argued that this at least suggests that something, presumably competition, has been keeping food prices better under control than the prices of holidays, fuel, housing, motor cars, computers and council tax. So far so good, but a glance at some of the constituent parts of the RPI rather spoils the effect. Nearly all the 8% rise in the 'all items' RPI since

Price inflation by main area 2000–2006
Retail price index (rebased from January 1987 to January 2000 = 100)

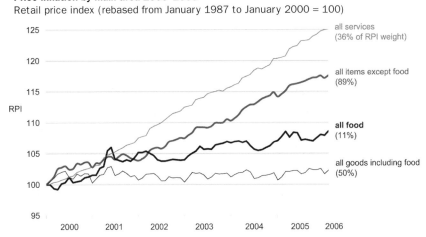

National Statistics, 2006

2000 has been produced by very steep rises in the prices for services. This should not surprise as most services share two characteristics that set them apart from goods.

First, few services are as exposed to global competition as most manufactures. You can have an electric kettle made in Shanghai by workers earning less than a twentieth of British wages and sell it on the UK market in competition with homegrown products. This makes it difficult for UK manufacturers to raise the price of kettles. The tea shop that uses the kettle to make tea for its customers is less directly exposed to global competition and so is that much less likely to exercise price restraint. There are bits and pieces of the service economy that do compete internationally, like airlines, some banking services and an increasing number of call centres and internet services, but they remain a minority.

Second, nearly all services are labour-intensive and labour has tended to be the fastest rising cost item, except recently for energy.

Food prices have risen much more steeply than the price of goods in general (including food) since 2000. Indeed non-food goods have even declined slightly in price. Of course, this is not necessarily to be blamed exclusively on supermarkets. It could equally well be taken as proof of the efficacy of the EU's Common Agricultural Policy in shielding our stricken farmers from the full blast of international competition, which would otherwise have brought rural incomes in the UK down to the level of the Horn of Africa.

How then can we best protect customers and the unmatched reputation of our supermarkets for value for money against a relentless campaign of vilification by Stalinist-inspired government interventionists aimed mainly at robbing shareholders of their just returns?

This example uses a relatively mathematically sophisticated value series to make the point. It takes the retail price index for food (usually based on 13 January 1987, which probably still has them grinning into their tea cups down at the Office for National Statistics, but recalibrated here to start in January 2000) and divides it by the RPI for all non-food goods and services, giving food price inflation as a percent of non-food price inflation. The result is a plausibly spikey curve that nonetheless declines fairly consistently from January 2000 to February 2006 suggesting that something, presumably competition among the big six food retailers, has caused this happy state of affairs. The fact that food prices have risen faster than the prices of all other goods is elegantly obscured. Two further touches raise this estimable chart far above the common rut.

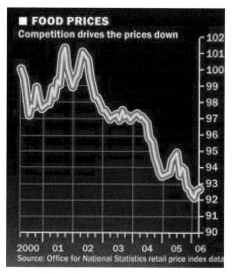

The Times, 2006

The headline 'Competition drives the prices down' – seemingly from the school that believes that when the rate of inflation falls, prices go down, rather than rising more slowly – clearly implies that food prices have fallen. The curve based on an abstruse value unit encourages this belief and seems at first glance to show a decline of 7% rather than the actual increase of 8% since 2000.

The interruption of the value axis between 0 and 90, and the choice of January 2000, rather than say January 2001, dramatises an otherwise unexciting optical impression.

The resulting PDQ is around 10:1 in the strict definition of optics:data (thanks mainly to the broken axis) and arguably around 20:1 if the inadequacy of the data itself is also taken into account. STD is close to zero and may even be negative, given the high-sounding seriousness of the apparently scientific y-axis. Graphic adjustment of reality doesn't come much better than this and it bears the hallmark of all truly great graphics, that it's impossible to tell whether the distortion is by accident or design. It is a fitting conclusion to the possibilities in category definition in general.

Having explored all the things we can do with the data for the category axis, it is now time to turn to manipulation of the category axis on the chart itself and what we can do to manage its relationship to the data constructively.

6 DISTORTING CATEGORIES AND TIME

Graphical Manipulation

Category definition and selection in the pre-graphical phase of communication offer varied manipulation opportunities. But once we get to designing the chart itself category distortion opportunities are even more attractive.

Manipulating category sequence

Telephone book sequence

The difference between a data table and a chart is that a chart has a message, a focus of attention, and a table hasn't. A telephone book, for example, is a rather grand data table, even in yellow pages form with thousands of subdivisions for the various commercial categories. The give away in both cases is the alphabetical order of the data categories, which is the almost universal default layout for data collections from which we want to extract relevant data quickly. As the differential probability of subscribers looking for one name rather than another is negligible, alphabetical order is acceptable except for those few numbers we might need very quickly indeed – fire brigade, police, poisons hotline and so on – which tend to get repeated on the cover or in the first few pages. This ought to mean that finding data categories in alphabetical order on a chart should alert us to the fact that the author has nothing particular to say. But, fortunately, things are a little more complicated.

This example goes back to an old maritime disaster, whose key ingredients of sex, age and class consciousness have kept it continuously in the public eye for 90 years. *RMS Titanic*, with 2,201 people on board, sank in the unusually calm but cold North Atlantic in the early hours of 14 April 1912 having hit an iceberg some two and a half hours earlier. *The Titanic* had enough lifeboats for 1,178; all were launched. Yet only 711 passengers and crew survived; 1,490 died (including 161 women and children – the traditional priority groups), mainly of exposure, before rescue ships arrived at dawn.

The primary causes of the disaster seem to have been faulty design: inadequate bulkheads, poor riveting and steel quality, lack of lifeboats and poor seamanship: reckless speed, poor lookout, failure of nearby ships to help. But attention has always focused on the human factors that allegedly unnecessarily reduced survival rates on the night: the untrained and ill-disciplined crew and incompetent officers, the segregation of third class passengers, other forms of class discrimination and the hysterical egotism of female survivors, who flatly refused to help those already in the water.

Fate of passengers and crew in the Titanic disaster, 1912
Absolute numbers by sex, age & status (traditional priority groups)

Group	Fate	Passengers by ticketclass			Crew
		1st	2nd	3rd	
men	died	118	154	387	670
	survived	67	14	76	102
boys	died	0	0	35	
	survived	5	11	13	
women	died	4	13	89	3
	survived	140	80	70	20
girls	died	0	0	17	
	survived	1	13	14	

Lifeboat capacity: 1,176

Wikipedia, 2006

The quantitative basis for most of the obsessive discussion since 1912 can be summarised in a simple eight by four data table. However, it's very difficult to use even this small quantity of information without going to the additional trouble of calculating totals and subtotals and horizontal and vertical percentage figures. A relatively new or, more precisely, rediscovered chart-type – the matrix chart – can both summarise this data and reveal the main relationships. For graphical connoisseurs this is an outstanding chart because it guides us unobtrusively through several layers of quantitative understanding. It makes for example five successive points almost at a glance:

Total 1st class survival rates were highest…

…largely because 1st class had the highest proportion of women and children…

…but nonetheless 1st class men had a strangely high survival rate…

…especially compared to 3rd class children…

…and, even more shockingly*, compared to 3rd class girls…

and catapults us ahead of some *Titanic* enthusiasts, who've been discussing these very points for decades. In other words it seems to be an entirely unfocused chart, in the sense of covering all known quantitative data in some detail, yet with considerable narrative power.

Fate of passengers and crew in the Titanic disaster, 1912
Absolute numbers by sex, age & status

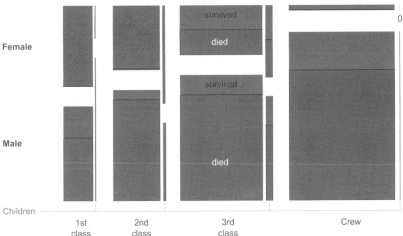

Data from: Dawson, *The 'unusual episode' data revisited,* 1995, & Friendly, *Extending Mosaic Displays,* 1999

Note that the sequence of categories on the x-axis of the chart is what a modern computer would choose by default. It also corresponds not only to what Edwardians would have considered rank or status (declining from left to right) but also to increasing male mortality rates. So the lack of conflict between arbitrary and focused category axis sequence is really just the result of the happy chance that they coincide in this particular example.

*by the widely accepted standards of the time

More than ninety per cent of Indian respondents plan to increase innovation investment
How will your company's investments in innovation compare with 2004?

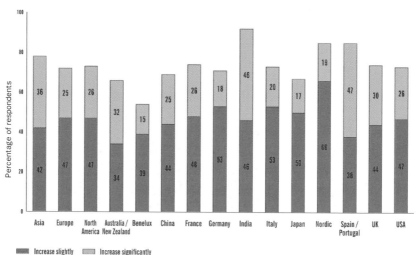

BCG, *Innovation*, 2005

When telephone book sequence does not coincide with the sequence necessary to support the point of the chart, the result is usually a weak chart that cries out either to be turned into a neutral table or into a chart with properly ordered categories. In this example from the Boston Consulting Group, India, the only country named in the action title of the chart is less than central to the message of the surrounding text (not shown here). The chart bears all the signs of having been prepared as background material to the pamphlet in general and hurriedly fitted out with an action title on the basis of the least unobvious conclusion to be drawn from its jumbled data.

Were the plans of Indian companies to increase innovation investment really of paramount importance to the message of the chart and its surrounding text, a version redrawn to separate the individual countries from the three main regional groupings and to display the countries in order of innovative intention makes the point more strongly.

A more interesting example of the useful ability of meaningless, telephone book sequence to stifle thought is provided by Niall Ferguson's book *The Cash Nexus*. In 'Dead Weights and Tax-Eaters' he argues that European welfare states limit 'deep relative poverty' to less than 5% of families, whereas former British colonies tolerate higher, the USA in particular much higher, proportions even after tax and transfers.

More than ninety per cent of Indian respondents plan to increase innovation investment
How will your company's investments in innovation compare with 2004?

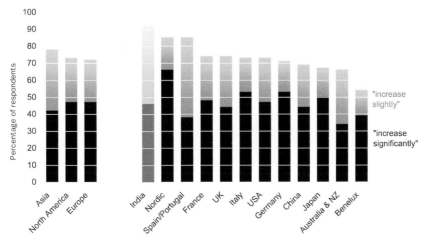

Data from: BCG, *Innovation*, 2005

On his figures this appears to be true, but is triumphantly obscured by a rigidly A-Z approach to graphical visualisation. Re-sorting the country axis by residual post tax and transfer levels of poverty makes the point about the ex-colonies very much more clearly.

It also strongly suggests that the European approach is deliberate, in the sense that the size of the adjustment of relative poverty figures by tax and transfer payment is determined more by the goal he hypothesises – 'less than 5% of the population' – than by the very widely differing pre-tax and transfer poverty rates themselves. The European instinct seems to be to do whatever it takes to get below the 5% limit. The redrawn chart also reveals with painful clarity that whatever robustly self-sufficient attitudes to social welfare the British may have once passed on to their colonies, the modern UK is firmly in the heart of the European camp, alongside other fabled haunts of Marxist-Leninist ideology such as Switzerland and Ireland.

To put it another way, whereas the redrawn chart openly provokes creative thought, the original would be better employed as part of an admissions test for aspiring telephone operators.

Relative poverty rates before and after taxation and transfers, 1991
Note: Poverty rates are defined as percentage of families with an income less than 40 per cent of the median

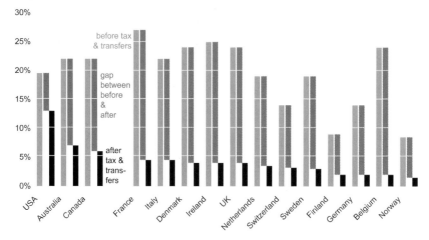

Ferguson, *The Cash Nexus*, 2001

In the face of such confusion of mind about the purpose of charts (as opposed to data tables) in general and of the message of these charts in particular, there seems little point in awarding PDQ (potential deceit quotient) and STD (sore thumb discount) values. Let's just file the technique under 'complicated ways of shooting yourself graphically in the foot'.

Explode pie at wrong point

Arbitrary order of the segments and categories in pie charts is just a variant on the more general telephone book technique of distraction. Readers and viewers of graphics are at least as hide-bound as the rest of humanity and expect pie chart segments to start at the 12 o'clock position with the biggest, continue clockwise in order of declining share and end with 'others' completing the full circle. Any deviation from tradition, as in this example, tends to distract, while readers search for some special reason for the unusual order of segments. None such is detectable here, in what is – for other reasons too – a spectacularly useless chart. The most obviously interesting geographical question about drugs is dependency or use per head of population; the second most interesting probably the relative popularity of different drugs in different parts of the world. This chart answers neither, but shows instead that the

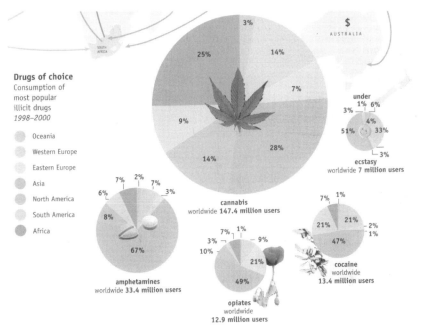

Drugs of choice
Consumption of
most popular
illicit drugs
1998–2000

- Oceania
- Western Europe
- Eastern Europe
- Asia
- North America
- South America
- Africa

cannabis
worldwide **147.4 million users**

amphetamines
worldwide **33.4 million users**

opiates
worldwide
12.9 million users

ecstasy
worldwide **7 million users**

cocaine
worldwide
13.4 million users

Smith, *State of the World Atlas*, 2003

distribution of marihuana use is somewhere between the distributions of world population and world GDP. Who'd have guessed?

But pies do lend themselves to their own special technique of optical distraction: the creative misuse of the click and drag facility to pull out, and thus stress, the least important segments of a chart. The interesting thing from our point of view about this technique is its rapidly increasing popularity, which promises quite soon to reduce its STD almost to zero and even to lead to its inclusion in the canon of officially neutral devices. Its origins are lost in the mists of time but probably go back to efforts to show very small shares at the resolution limits of the medium being used. But nowadays they are often to be found in any pie chart with typographical layout difficulties. In this case the technique has been used on two pie charts marooned in the Indian Ocean purporting to show the number of refugees in India and Tanzania. Showing that 3,000 of the almost half a million refugees in Tanzania are from Somalia is perhaps justifiable as a 'straw-in-the-wind' statement about the Horn of Africa, but stressing that 300 out of some 345,800 refugees in India are neither from Sri Lanka, Tibet, Burma, Bhutan, Bangladesh nor Afghanistan is both useless and and raises quite unjustified claims to precision.

Refugees

More than 40 million people who have fled war or repression are unable to return to their homes; many have been made refugees more than once.

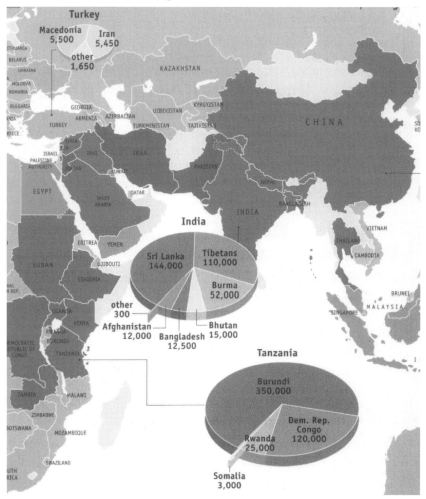

Smith, *State of the World Atlas*, 2003

The final example of misplaced and hence distracting emphasis of pie segments is taken from the German version of the *State of the World Atlas*. It is misbegotten from several points of view, not least in presenting figures for which it is difficult to think of a useful purpose. It is basically an account of where over-75-year-olds live, which corresponds to where most people in general live, tempered only by the much more interesting but missing statistic of the proportion of each population over 75. The chart thus rather elegantly concentrates exclusively on what most people don't want to know, without committing the vulgarity of making this explicit in its title.

Star billing, in the middle at the top, is given to the 2.1% of 75-year-olds, who live in that neglected superpower, 'rest of the world'. The other segments of the pie projecting beyond the circumference represent the four next smallest groups. Against this background, the random sequence of segments (even the old trick of spelling them in French and then seeing if alphabetical order fits, which often works with EU publications, provides no enlightenment here) is just a strategic distortion reserve.

Older than 75
Share of the world population over 75 in %

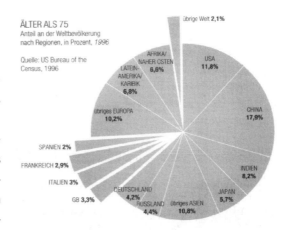

Smith, *Der Fischer Atlas zur Lage der Welt*, 1999

Manipulating x-axis scaling

Stretch/shrink to change gradient

Arbitrary category sequence and misplaced pie chart emphasis lead to general confusion and weaken messages. Although this can be used for quite deliberate and targeted deceit, manipulation of the category axis only really comes into its own with techniques that bend the relationship between the data and the optics in a more calculated way. Many of these techniques are just twins of similar ruses on the value axis, but are none the less powerful for that.

The most popular is also the simplest and involves stretching or shrinking the time axis irregularly to alter the actual gradient of a curve or the virtual gradient of a row of bars or columns.

Management consultants depend heavily on change or, more precisely, on the fear of change among their potential clients. So anything, be it but a thimbleful, that tends to foster the impression of faster change in a threatening world still swells the mill race of management consultants' marketing efforts. In this case McKinsey was trying to persuade us that the Chinese car market, after a faltering start in its first hundred years, had just achieved lift-off and was destined to roar virtually out of sight in the near future. Anyone with ambitions to sell to, or even buy from, the Chinese car industry would clearly be well-advised to avail themselves of the services of the consultant that spotted this important trend.

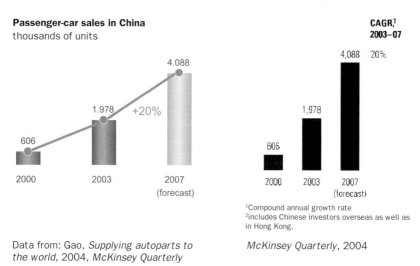

Passenger-car sales in China
thousands of units

CAGR,[1]
2003–07

Data from: Gao, *Supplying autoparts to the world*, 2004, *McKinsey Quarterly*

McKinsey Quarterly, 2004

[1]Compound annual growth rate
[2]includes Chinese investors overseas as well as in Hong Kong.

McKinsey's interpretation may well have been right but seems to have faced an irritating little presentational problem. There must have been at least one gap in the supporting Chinese data, of which only 2000, 2003 and a forecast for 2007 are shown in the chart. The obvious way to display this data would have been to use the neutral layout shown on the left, either with columns or a curve or even with both, to emphasise both absolute size and the linear nature of the impressive growth rate.

Luckily for connoisseurs of the not quite untrue, McKinsey was not so craven and chose instead (above right) an optically regularly spaced time axis with a clearly – alright, charmingly discretely – marked break in the axis between 2003 and 2007.

Those of our ancestors who were unable to spot the difference between a straight line and bent line are not in fact our ancestors at all, having departed this life and our gene pool several million years ago when misjudging the position of a neighbouring branch towards which they thought they were leaping. So the lines joining the columns in the third and simplified version of this chart are only there to reinforce the point that the McKinsey version does rather more than full justice to the expected growth of the Chinese car market and no disservice to its publisher's marketing prospects in the process.

The PDQ involved is modest, around 1.2:1, but STD is as close to zero as makes no difference. From our point of view it's a pity that mere self-respect seems to have driven McKinsey to include that squiggle to mark the broken axis, as its presence alerts the reader to the break and to the fact that there was no very convincing graphical

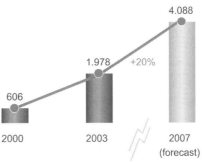

Data from: Gao, *Supplying autoparts to the world*, 2004, *McKinsey Quarterly*

necessity for it to begin with. Nonetheless, the misleading impression created by the finished chart, for all but the most leisured reader, is circumspect in its ambitions, elegantly achieved, entirely deniable and thus an ornament to the management consulting profession in general.

The technique is surprisingly popular and is often used in contexts that camouflage it even better than the example above. In its depiction of the numbers attending the London Gay Pride March in 1987, 1993 and 1997, the *Penguin Atlas of Human Sexual Behavior* spaces the years equally, although, of course, the gap between 1987 and 1993 should be two-thirds bigger than the gap between 1993 and 1997. This tends to obscure the relatively slow take-off of the event and the fact that by 1997 saturation point seems to be being approached. Organisers of

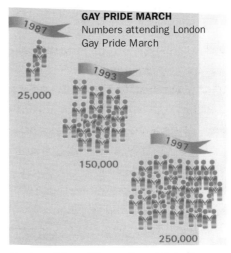

Mackay, *Atlas of Human Sexual Behavior*, 2000

the 1998 March only claimed 50,000 attended, the police 17,500, but there may be considerable definitional problems, which put mere category axis spacing in the shade.

The technique turned up recently, shorn of any disclaimer or axis break warning in the usually fastidious *Handelsblatt*, the German equivalent of the *Financial Times*, on the subject of the employment quota of Japanese women since 1965. The example is interesting mainly because it illustrates an all too human error that the improving availability and frequency of official statistics tends to encourage and which we can turn to our own account.

Employment quota of Japanese women %

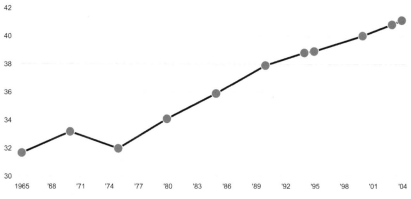

Data from: *Handelsblatt*, 2005

A neutral chart shows a fairly steady rise in the proportion of Japanese women working, with a slight dent in the early 1970s and an almost imperceptible slowing in the rate of increase since 1990.

The *Handelsblatt* original, on the other hand, seems to betray an over hasty transfer from the original spreadsheet – in which all available years were probably lined up side by side without gaps – to the chart drawing function itself, without taking the trouble to insert dummy columns in the spreadsheet to represent the years for which data was missing. The difference between the two renderings is small, but the neutral version above suggests that the increase in the employment quota has quickened slightly in the last ten years, the lower version that it has slowed.

To sum up, distortion of the category or time axis seldom offers very large PDQ values – 1.2 or 1.3:1 are respectable results – because STD

Climbing slowly
Number of Japanese women in the workforce (millions)

Langsam nach oben

Berufstätigkeit japanischer Frauen

40 *Anteil der Frauen
an den Erwerbstätigen
in Prozent*

41,1 2004
40,8 2003
40,0 2000
38,9 1995
38,8 1994
37,9 1990
35,9 1985
34,1 1980
33,2 1970
35
32,0 1975
31,7 1970
1965
30

Handelsblatt, 2005

values start to rise so sharply as the axis is differentially stretched, like cheap chewing gum, that the point of the manipulation is all too easily lost.

Split the category axis

With the possible exception of the Congregation for the Doctrine of the Faith (or Inquisition, as it used to be known), no human institution has, since its inception or before, been the subject of such relentless criticism as the National Health Service. The public does not take kindly to the application of economic principles to the care of the sick, still less to the prevention or early detection of sickness. So it's scarcely surprising that comment on the NHS offers many exciting examples of the graphical adjustment of reality to fit whatever case is being made.

Take the question of how well breast cancer screening programmes correspond to the needs of women and the capabilities of the programmes themselves to prolong life significantly. A neutral comparison of the coverage rates of screening and of cancer incidence by age group seems at first glance to reveal some strange anomalies. While maximum coverage (defined here as the proportion of women in the age group to have been screened within the last three years) coincides quite well with initial peak incidence between the ages of 55 and 64, coverage of the over 75s and of the 45–49s is very low although cancer incidence in these groups is far from insignificant. However, upon reflection, the chances of over-75s dying of something else quite soon are high enough

UK breast cancer incidence and screening coverage by age group 2002–2003

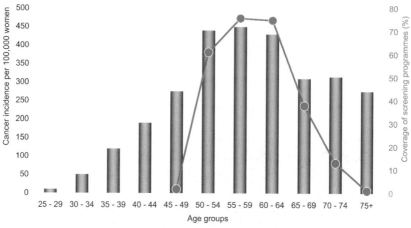

Data from: *The Times*, 2005 & Cancer Research UK, 2005

to make expensive screening uneconomic. This is not true of the under-50s. In other words, there are considerable gains in clarity from putting screening coverage and breast cancer incidence above one category axis.

The information content of *The Times* version below is approximately the same, but the comparison of coverage and incidence requires considerable mental gymnastics and the relevant questions are more likely to be avoided, especially as the two category axes are not only separated but also have entirely different spans and intervals.

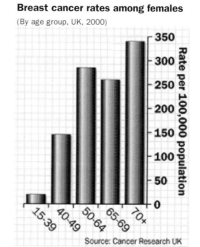

The Times, 2005

Inappropriate equal spaced scale

In the examples so far, the manipulation of the category scale has depended on selective stretching of parts of the axis or even on the omission of crucial categories like intermediate years. But there are also many cases in which the opposite manipulation opportunity lies in using constant category intervals.

Distribution of Wives by Race in Married-Couple Families by Race of Husband 1992

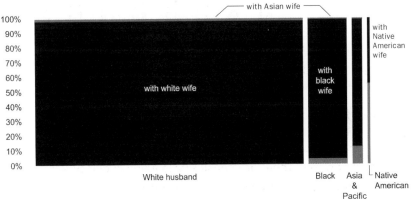

Data from: US Bureau of the Census, *Fertility of American Men*, 1996

The trouble with melting pots is that they make such a mess when they boil over. Most reasonable people are therefore at pains to prevent boil-over, but this admirable ambition can lead to some highly original graphics of undoubted power but uncertain effect. Take for example the knotty question of the degree of racial integration in the USA. The US Bureau of the Census, a well-known haunt of extremely reasonable people, laid the facts before the public in a Technical Working Paper in 1996. One of the key statistics, not surprisingly, was the prevalence of intermarriage among the races. In 1992 the slightly depressing facts were these, showing that intermarriage was still very, very unusual. The width of the columns corresponds to the number of marriages in the entire USA by the race of the husband and thus the area of each segment is proportional to that type of marriage's share of all US marriages. Only the tiny red areas represent mixed-race marriages. The overwhelming black areas show the proportion of same race marriages. To put it another way, intermarriage in the USA in 1992 was so rare, that displaying it still challenges the optical resolution of our printing technology to this day.

How can we soften the blow and present the figures in a slightly less depressing way? One way is to variagate the overwhelming blackness of the single race marriages by assigning shadings to the wives by their actual race rather than simply by whether it is the same as the husband's. This makes quite an improvement, though we are still stuck with a dominating block of white men married to white women, which takes up four fifths of all areas.

Distribution of Wives by Race in Married-Couple Families by Race of Husband 1992

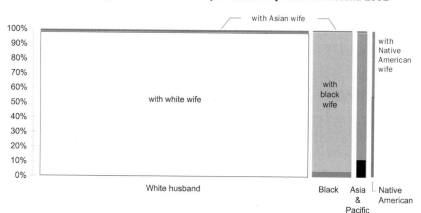

Data from: US Bureau of the Census, *Fertility of American Men*, 1996

But this too can be solved graphically if we redraw the category axis to give equal prominence to all races. And who knows, it might even earn us a mention in dispatches in a *Guardian* editorial. Whatever the reason, it was in fact the format eventually chosen by the US Bureau of the Census.

The result is a simperingly harmless chart, that requires the help of a pocket calculator and a working knowledge of the racial structure of the US population to an accuracy of a couple of percentage points before it reveals any useful information at all. And yet this admirable obfuscation has been achieved not only without telling a single lie, but also by adhering strictly to the most conservative and neutral graphical standards. Which suggests that there is either something wrong about not telling lies, a subject on which this book could not possibly comment, or about adhering strictly to conservative graphical standards: upon reflection, probably the latter.

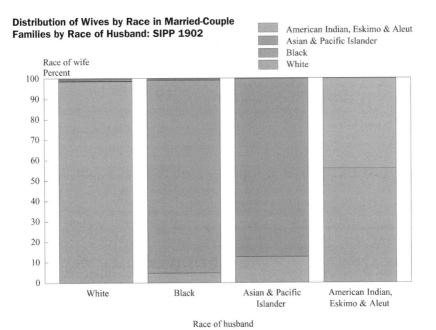

Distribution of Wives by Race in Married-Couple Families by Race of Husband: SIPP 1902

American Indian, Eskimo & Aleut
Asian & Pacific Islander
Black
White

Race of wife
Percent

Race of husband

US Bureau of the Census, *Fertility of American Men*, 1996

Arbitrary x-axis scaling in 3D

Few human activities give rise to such obsessive quantification as sport. Wisden for instance makes the Proceedings of the Royal Statistical Society read like a Barbara Cartland novel. And where there's a numerical series, some interesting graphical distortion can't be far behind.

For fans of the 100m sprint, the question of the ultimate limit of human performance is particularly fascinating and the history of our approach to that limit, wherever it is, has been unusually precisely documented since William McClaren of the UK won the first official world 100m record in 1867 with a time of 11.0 seconds. The record stood for an amazing 45 years. Modern records are less long lived, although this is partly to do with the fact that they have been recorded with an accuracy of 0.01 seconds (as opposed to the previous 0.1 seconds) since 1968. For devotees of the absolutely meaningless factette, this enabled Jim Hines to break his own record of 9.9 seconds, set in the previous June, with a time of 9.95 seconds, on the face of it 0.05 seconds slower, in October 1968.

Faster and faster – how the world 100m record has fallen since 1867

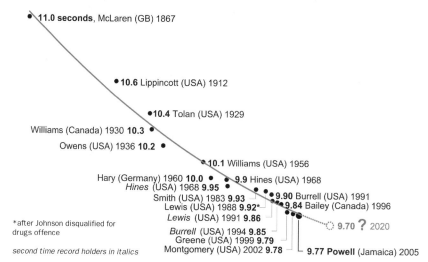

● 11.0 seconds, McLaren (GB) 1867

● 10.6 Lippincott (USA) 1912

● 10.4 Tolan (USA) 1929

Williams (Canada) 1930 10.3 ●

Owens (USA) 1936 10.2 ●

● 10.1 Williams (USA) 1956

Hary (Germany) 1960 10.0 ●   ● 9.9 Hines (USA) 1968
Hines (USA) 1968 9.95 ●
Smith (USA) 1983 9.93 ─┐   ● 9.90 Burrell (USA) 1991
Lewis (USA) 1988 9.92* ─┘   ● 9.84 Bailey (Canada) 1996
Lewis (USA) 1991 9.86 ─┐

*after Johnson disqualified for drugs offence   ⟨⟩ 9.70 ? 2020

second time record holders in italics

Burrell (USA) 1994 9.85 ─┐
Greene (USA) 1999 9.79 ─┘
Montgomery (USA) 2002 9.78 ─┘ └ 9.77 **Powell** (Jamaica) 2005

Data from: *The Times,* 2005 & Deutsche Leichtathletik Verband, 2003

Very slightly more seriously, the decrease in world 100m record times seems to have accelerated since the mid-1980s, but it's difficult to believe that this trend will continue unless recent genetic advances soon reach the track, resulting in the creation of a new breed of athletes consisting only of lungs, buttocks, thigh muscles and, presumably, drastically redesigned Reeboks. Come to think of it though, traditional training methods have already brought us quite close to this goal.

But what if you are a sports journalist and are worried that this sort of idle speculation will fail to grip the modern reader? You have two possible remedies. One is to switch on BBC Radio 4 long wave during the cricket season and let even idler speculation wash soporifically but reassuringly over you, and the other is to get the graphics department to jazz things up a bit.

The Times seems to have chosen the second option and in the process achieved the difficult feat of scaling both axes roughly proportionately to the world record times themselves. This has the interesting side effect of destroying the time axis and thus any possibility of pondering the probable future development of world record times for the 100m. Add to this some chart junk for the benefit of those readers who have difficulty recalling what athletics tracks, athletes and stopwatches

The Times, 2005

Men's world 100m records (seconds) 1960–2004

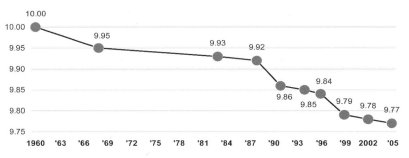

Data from: *The Times*, 2005

look like, and the result is a deeply gratifying yet puzzling chart (why, for example, is the track floating about five inches off the ground? Why have the sprinters left their lanes? Will the man in front be disqualified for blocking others? Isn't the man in sixth position doing suspiciously well for someone with one leg?) that cries out to be returned to the data table from which it was untimely ripped.

The relevant portion of the neutral chart, for simplicity's sake without the names of the record holders, has been included below the enlivened 3D version to demonstrate how neatly data has been lost by the suppression of the time axis.

7 DISTORTING THE WHOLE CHART

Mismatch Title and Data

So far we have been looking at deceit techniques that are applied to one or other of the usual axes in a typical quantitative graphic. But there are large numbers of techniques that cannot be attributed to any one particular axis. In these cases, the point is either dislocation between the purported message of the chart, often contained in an 'action title' but otherwise derived from the surrounding text, and the original data behind the graphic. In other words, the problem lies in the fact that the chart doesn't prove what it claims to prove.

Obviously there are as many variants of this type of deceit as there are errors of logic and failures of empirical proof in published statements of any kind. Once again our field of choice is invigoratingly wide, but we can group these possibilities into two main groups according to the point in the production of the graphic at which the deceit occurs:

1 During data collection and preparation, including:
   - mismatch between message and data
   - exaggeration of the reliability and precision of the data
2 During the drawing of the graphic itself, including:
   - graphics for graphics' sake
   - distorting layout and illegibility
   - choice of the wrong chart type

Mismatching title and data

Simple contradiction

In the days when charts were produced by recalcitrant serfs toiling in the cellars for their betters in the panelled boardrooms above, most of the lines and lettering on the charts were stuck on with double-side tape or even spray adhesive. So the chances of an action title sticking to the wrong picture, or even two charts becoming mysteriously merged, were quite good. An action title comically at odds with the body of the chart was just as much an occupational hazard for chart-churning consultants as missing teeth for ice hockey stars or insanity for hatters. But in the new, glueless age of DIY computer graphics, the scope for this sort of organisational or technical error has narrowed and we are thereby deprived of a convenient method of excusing gross graphical incompetence. But now and again you do find yourself involuntarily scanning the rest of the page in the hope of finding the white space from which, in that bygone age, an exceptionally inapposite action title could have become unglued.

A case in point is this *Sunday Times* contribution to the debate about possible UK membership of Euroland, under the action title 'The euro hasn't been good for employment...' Note that civilized adults can reasonably differ about whether this statement is true or not. The more modest question here only concerns the quality of the link between the chart and its conclusion.

At first sight, indeed at second or third sight, it is very difficult to find anything in the body of the chart to support the title. According to the chart, unemployment in France and most of the rest of Euroland was already falling by the time the euro was launched at the end of 1998 and the improvement quickened slightly in the two years after that. But rates started to increase again at the end of 2000, a year before the introduction of euro notes and coins, and have risen further since, though to levels shown here below those before the launch of the euro, let alone those at the time of the announcement of the euro timetable in 1996.

Perhaps the *Sunday Times* was simply the victim of its own honesty in displaying such generous historical perspective. If we were to delete all time periods before July 2001 and change the label 'Switch to euro notes and coins' to, say, 'Real euro launch', we could get chart and conclusion back into synch without very many readers noticing. Courtesy forbids more than cursory mention of an alternative explanation, which would be that whoever was responsible for this chart genuinely thought that the economic effects of a currency only start when its notes and coins enter circulation.

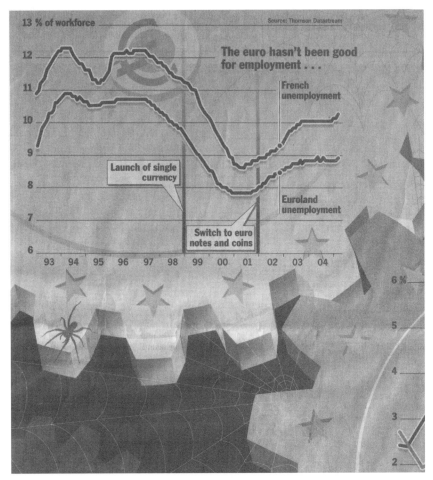

The euro hasn't been good for employment . . .

French unemployment

Launch of single currency

Euroland unemployment

Switch to euro notes and coins

Source: Thomson Datastream

Sunday Times, 2005

The lesson for practising deceivers is that the chances of being exposed for this sort of shenanigans are vanishingly small. Few will adduce the counter-arguments above and fewer still will murmur 'post hoc ergo propter hoc?' ('After this, therefore because of this?'), which is the most powerful objection to the whole sorry mess.

Mathematical misunderstanding

If the most charitable explanation of the last example lies in a certain lack of economic sophistication, innumeracy is fortunately almost as common, as this example from the BBC's website suggests, and can also be harnessed to our requirements.

This book is not about to get involved in such a complex and, as presented here, poorly defined comparison of the costs of different modes of transport. The question is merely whether, as claimed by the BBC, 'This chart shows how public transport costs are actually greater than the cost of running a car'. It doesn't. It only shows that the costs of rail and bus transport have grown faster than that of cars since 1987. Which costs *are* actually greater depends on their relative size in 1987 as well as subsequent changes.

Car vs Public Transport

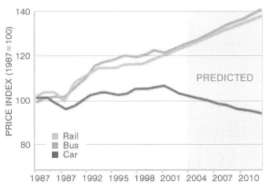

Costs compared

If current trends continue, the cost of motoring is forecast to fail by a further 20% in the next 10 years.

However, this chart shows how public transport costs are actually greater than the cost of running a car.

Complaints about the cost of fuel are usually linked to the perceived high overall cost of motoring.

BBC, 2006

Again, this technique is relatively robust and will be spotted by few. The only drawback is that the few who do notice will find it difficult ever to take you mathematically seriously again.

Sloppy language

The chart overleaf from the *Penguin Atlas of Human Sexuality* was last seen as an example of how to distort the value axis in order to minimise differences between categories. It's playing a return fixture here because it also shows how easy it is to smuggle conclusions past an audience with the use of sloppy language. The small print of the label on the left reads 'Average age of children receiving sex education', which would make the UK figure of 11.4 years fairly impressive. After a little thought, the context makes fairly clear that it's the average age of *first* sex education that is being displayed. This would put the 'average age of children receiving sex education' somewhere beyond 14 or even 15.

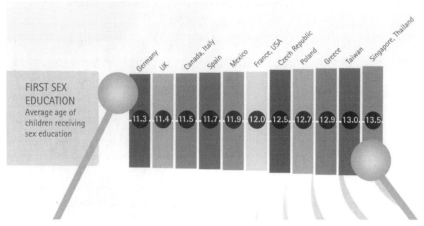

Mackay, *Atlas of Human Sexual Behavior*, 2000

Simple exaggeration

It's always difficult to know quite what to make of phrases like 'the world's strongest currency'. Forget such nitpicking details as whether the world really only has four currencies worthy of note, which might raise an eyebrow or two in Shanghai, Mumbai and Rio. Forget too that, as always with indices, it all depends where they start and what the position was at that time. Had the pound in 1997 been in some sense significantly 'weaker' than the yen, or even than the still unlaunched euro, the curves seem to show that the same would have been true in early 2005 as the index was still very close to 100 in both cases.

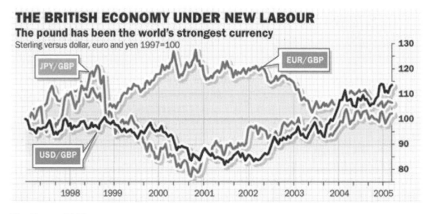

The Times, 2005

Even on its own eccentric terms, the chart appears to show the pound having been almost uninterruptedly 'weaker' than, say, the dollar from 1997 to the end of 2003 and the yen from early 1999. Were the suggestion not implausibly machiavellian, one could deconstruct the fatuous action title of this chart as a deliberate attempt to make New Labour look faintly ridiculous on the eve of the general election in the spring of 2005: 'We don't want to fight but, by Jingo, if we do, we've got the ships, we've got the men and we've got the world's strongest currency too.'

Non sequitur

The fact that a statement is true doesn't necessarily mean that the argument upon which is based or the chart of which it forms the action title is itself sound.

In this example from the *Atlas of Women in the World*, the argument seems to run something like this:

– All girls infected with HIV were infected by males.
– But boys' infection rates are lower than girls'.
– Therefore, many girls were infected by older men.

Except for the 'therefore' it's quite likely that all three statements are true or nearly so. But to justify the 'therefore', without which the chart's title is simply out of step with the facts displayed, we need two other statements to be equally true:

– Male to female infection is as likely as female to male.
– Boys and girls are similarly promiscuous.

Strictly speaking the fact that neither of these statements is true in African or most other human societies is beside the point. The problem is that the logical link between the action title and the data displayed is weak to the point of caricature. If, to take an absurdly extreme example, all girls only had one and the same sexual partner, who was the only infected boy in town, boys' infection rates would be negligible but females' limited only by the probability of transmission. On its own the ratio of female to male infection rates proves nothing at all.

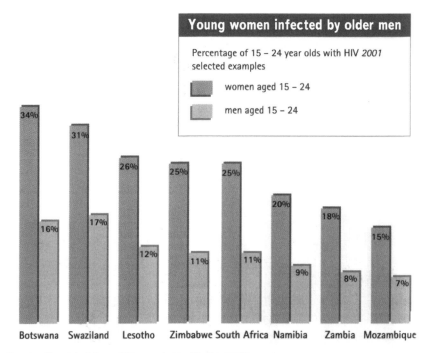

Seager, *Penguin Atlas of Women in the World*, 2003

Note that the figures in this chart are not at issue. Irrelevant though they are, they do provide a powerful backdrop to the message at the top of the picture and it is unlikely that many readers will pause to consider the logical gap between the medium and the message.

Implausible correlation

Chaos theory, indeed most of human experience, predicts that the smallest cause can have the greatest consequences. As the imaginary headline has it: 'Franz-Ferdinand found alive, two World Wars annulled'. But there are limits, even in graphics.

WirtschaftsWoche is an old established and successful German business weekly not noted for an overly green approach to the world's problems. Indeed there is some doubt about the direction of the causation that *WirtschaftsWoche* seems to be suggesting here. Is it the extra air conditioners increasing vehicle fuel consumption and thus climate changing emissions, or is it the emissions causing global warming and hence encouraging perspiring Germans to buy cars with air conditioners?

In either direction the suggestion is a trifle far fetched, whether in view of the vastly greater contribution to emissions made by many other factors or of the very small changes in average temperatures so far observed.

Increasing burden
Car with air conditioners in % of new registrations (columns) & climate changing emissions (line, mill. tonnes of CO_2 equivalent)

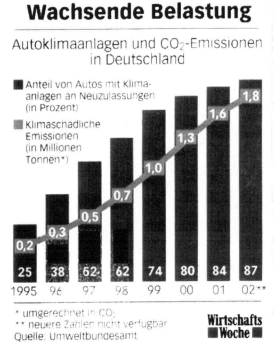

WirtschaftsWoche, 2005

While putting two data series on one chart runs the risk of being interpreted as a statement of causation, the implication is eminently deniable and so of considerable interest to those practising to deceive.

Pseudo correlation

If you're really desperate to find some sort of correlation to add respectability to an otherwise unimpressive train of thought, you can always turn to the old trick of using two variables that are separated only by a logical or mathematical constant. Sounds complicated? So much the better.

Stripping away the jargon from this Boston Consulting Group strategic matrix reveals that the two axes are simply 'recent growth in penetration' and 'present penetration' of imports into the USA from rapidly developing economies like China and India. Recent growth obviously tends to enhance present penetration and so the axes are scarcely mathematically independent.

RDE's Share of the US Market is gaining momentum

BCG, *Focus*, 2005

But then comes the rather daring analysis. High penetration, low recent growth is classed as 'moving early'; high penetration, high recent growth as, wait for it, 'growing fast'; low penetration, high recent growth as 'up and coming' and finally low penetration, low recent growth as 'globalising slowly'. Note that even the most hopeless case, where the Chinese toe is barely in the door and no great progress has been made for five years, still rejoices in the portentious title of 'globalising slowly', evoking a disembodied hand waiting to drag doe-eyed American industrialists into the crypt.

Against this background, the action title is ominous but still modest: 'RDE's share of the U.S. market is gaining momentum'. The implied acceleration is unsupported by the chart, which shows, at most, positive growth rates and varied present market penetration figures.

Exaggerating data precision and reliability

If we are not prepared to take a lot on trust, modern life quickly becomes unliveable. Indeed exaggerated distrust of others, whom we suspect of having planted miniaturised transmitters in our teeth or of being agents of extragalactic powers, is a good diagnostic of incipient schizophrenia. On the other hand, some recent high-profile scientific fraud and allegations of corruption against the most august in the land have heightened our awareness of human frailty. The internet has also brought more of us into direct contact with criminals than the Post Office ever achieved. Who these days has never been in contact, however passive, with Nigerians seeking the last £10,000 they need to collect a fortune or has never received an unctuously concerned email from the USA asking us for our account number and password to restore the function of bank cards supposedly cancelled to avoid fraud?

So, like everyone else, the audience for graphics has become less trusting and the premium for suggesting that our charts fulfil the highest standards of precision and reliability has grown. In the real world, precision and reliability are often inversely related; it is after all usually easier to be roughly than precisely right, but in the quaint world of quantitative visualisation the former is often mistaken for the latter. And so often overrated.

Statistical overconfidence

Surely someone who gives you a figure to several decimal places must be surer of the facts, perhaps more honest even, than someone who tries to flit by on a wing and an integer? True or not, this prejudice has been cluttering up charts for generations. Take for example the results of a survey conducted by the US Bureau of the Census into internet use in the USA.

The Bureau of the Census is both scrupulous and informative in its use of statistics and provided several pages of formulae, tables and explanations to help readers judge the accuracy and confidence limits for these results. It turns out that the 90% confidence limits for that '41.3% of US population does not use the Internet' stretch from less than 41% to over 41.5%. This is an accuracy to be proud of and was achieved by a sample size of several million, reached by tacking the internet questions onto the annual labour force survey. Even so, the probability of it really being 41.3 rather than, say, 41.2, 41.4 or even 41.1 is not much bigger than our minimal concern about whether it is or not. Indeed this residual inaccuracy is only that which results from statistical sampling error, which can be calculated precisely, whereas non-sampling errors can only be roughly estimated and are probably

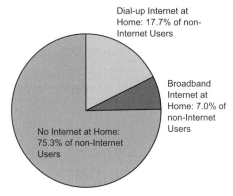

1.3% of US Population Does Not Use the Internet

Dial-up Internet at Home: 17.7% of non-Internet Users

Broadband Internet at Home: 7.0% of non-Internet Users

No Internet at Home: 75.3% of non-Internet Users

US Department of Commerce, *Entering the Broadband Age*, 2004

bigger. In other words, the survey shows a proportion of around 41% not using the Internet, and adding a decimal point is less empirical than inspirational.

Overprecision

The example from the Bureau of the Census is very mild and should be seen in the context of avoiding that irritating little message on data tables to the effect that the columns don't add up exactly to the totals because of rounding errors. But this example about the giants of the fish farming business goes further.

If you want to show national shares of world production, stressing the comparative insignificance of the German contribution, the standard method would be to use a pie chart, which makes the point simply but a bit boringly: China first with almost 28 million tonnes, the rest minnows and Germany barely visible with a mere 49,582 tonnes.

The graphical problem is showing 50 thousand and 28 million (560 times as much) on one chart. This may be an exceptional case where 3D makes sense.

Suppose you want to use a standard column chart and Excel has just asked you to choose a tick mark interval. On a '1D' column chart a reasonable answer – leaving a bit of space at the top – might be, say, 5 million tonnes needing six grid lines reaching to a maximum of 30 million. The German share would be beyond the limits of print resolution. In 3D, 560 little cubes representing 50,000 tonnes each could be arranged in 10 by 10 layers of 100 cubes. The Chinese stack would be 5 layers tall with an incomplete layer on the top (each full layer representing 5 million tonnes as in the 1D version) compared to a single cube for German production. 3D would in

Giants of fish farming 2002
Total world production: 40 million tonnes

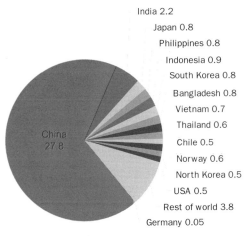

India 2.2
Japan 0.8
Philippines 0.8
Indonesia 0.9
South Korea 0.8
Bangladesh 0.8
Vietnam 0.7
Thailand 0.6
Chile 0.5
Norway 0.6
North Korea 0.5
USA 0.5
Rest of world 3.8
Germany 0.05

Data from: *Focus*, 2004

effect be allowing 100 times greater resolution. So we should catalogue this technique under useful ways of fitting big and small numbers on the same page while noting that it offers most of the deceit possibilities of any iconic graphical treatment in 3D.

It is also roughly the technique used by *Focus*, Germany's second largest news weekly, but only roughly. In fact, Focus chose cubes representing a startling 49,582 tonnes each, for the simple reason that this was the precise size of German fish farms' output in 2002. But

Giants of fish farming
Annual production in tonnes

Focus, 2004

rounding the German production quantity to 50,000 would only involve an inaccuracy of 0.3%, well within the inherent inaccuracy of this sort of statistic, corresponding to roughly one day's output per year. Rounding to the nearest 50,000 would involve an error in the case of China, by far the biggest producer, of 0.06%, scarcely an inaccuracy likely to alter the message of the chart. It would also almost halve the amount of ink used to display the figures.

The lunatic over-precision – suggesting that we can really be sure that Chinese production was only 27,767,251 tonnes and not, say, 27,767,252 – distracts from the message and even undermines the plausibility of the entire chart. Nice one!

Examples using even larger figures are only too easy to draw from the political arena. Nothing is as difficult to predict as the future and we would usually do well to avoid implausible precision when talking, let alone making promises, about it. But bringing together politicians' chronic need to make promises and their crumbling credibility with voters can produce a heady graphical mixture.

The debate, if so Socratic a term is justified, about future levels of government spending dominated much of the 2005 election campaign. In the process the English language was subjected to a sadistic wracking that would have made Torquemada – that fanatical Grand Inquisitor – pale at the gills. 'Cuts' for example, hitherto bordering on the blunt in its lack of ambiguity, came to mean successively 'rate of change', 'difference between our rate of change and theirs' and finally, and no less bluntly, 'increase'. But then they were talking about taxation.

Into this maelstrom of recrimination floated *The Times* with a sober comparison of Labour and Conservative spending plans for the next six years. Bearing in mind that the Treasury is modestly pleased if it manages to predict government spending over the same number of months with an error of less than a few hundred million, the precision of both parties' figures for, say, 2011–12 should leave us shaken but otherwise unstirred.

Two features are noteworthy. First, the figures for the Labour party's plans are, as a footnote unobtrusively makes clear, not direct quotes from party HQ but concocted by multiplying the Conservative GDP estimates by the Labour government's tax quota in 2006–7. This is daring but, in the interests of comparability, pardonable. Second, an extra decimal place in the Conservative figures for % of GDP in 2011–12 betrays the real reason for the implausible precision of their spending plans. The 40% barrier is breached. Assuming conscientious rounding techniques, it must have been by less than 0.005% of GDP but it's the thought, or more precisely the sound bite, that counts.

£35bn CUTS: WHAT THE TORIES PLAN...						
Labour will do						
	2006-07	07-08	08-09	09-10	10-11	11-12
£bn (*TME)	549.2	580	606.1	634.6	665.8	698.4
% GDP	42.1	42.4	42.2	42.1	42.1	42.1
The Conservatives will do						
£bn	541.7	567.4	589.8	612.3	638.3	663.5
% GDP	41.5	41.4	41.1	40.6	40.4	39.99
Tory cuts (£bn)	7.5	12.6	16.3	22.3	27.5	34.9

*TME figures after 2006-07 assume total spending remains constant as a share of GDP as assumed in the Tory for Money Action Plan

The Times, 2005

Spurious data reliability: over footnote

The literary equivalent of absurd numerical precision is over footnoting. This example is taken from Niall Ferguson's highly readable work of historical scholarship on the connection between money and power, *The Cash Nexus*.

The chart itself is an admirable example of the 'make the buggers work for it' school of graphical exposition. Even those readers in possession of keen eyesight, youthful neck muscles, a good memory for logarithmic tables, a drawing board and a sharp pencil will have difficulty agreeing with Professor Ferguson that Figure 4 shows anything, let alone with the hubristic proposition 'As Figure 4 shows, such levels [of military spending as a % of GNP] were rarely attained in the nineteenth century. Between 1850 and 1914 the highest proportion of GDP consumed by the British armed services was just 11 per cent …'. Until, that is, their eyes get as far as the intimidating footnote and the sheer impertinence of disagreement of any sort dawns upon them. This is less a footnote than a reading list for a BA. Doctoral theses have been built on less.

SPENDING AND TAXING

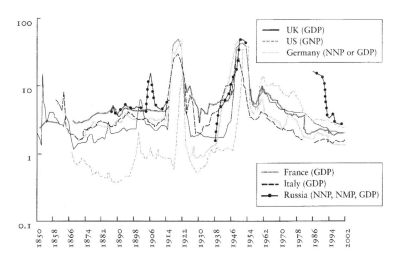

Figure 4. Defence spending as a percentage of national product, 1850–1998
(log. scale)

Sources: Defence spending: UK: 1850–1914: Correlates of War database; 1914–88: Butler and Butler, *British Political Facts*, pp. 393 f.; 1989–98: SIPRI. US: 1870–1913: Hobson, 'Military-extraction Gap and the Wary Titan', p. 501; 1914–85: Correlates of War database; 1986–98: SIPRI. Germany: 1872–1913, 1925–32: Andic and Veverka, 'Growth of Government Expenditure', p. 262; 1933–8: Overy, *War and Economy*, p. 203; 1938–44: Petzina, Abelshauser and Faust (eds.), *Sozialgeschichtliches Arbeitsbuch*, vol. iii, p. 149 (however, 1933–43 percentages are from Abelshauser, 'Germany', p. 138); 1950–80: Rytlewski (ed.), *Bundesrepublik in Zahlen*, pp. 183 f.; 1982–98: SIPRI. France: 1820–70: Flora *et al.*, *State, Economy and Society*, vol. i, pp. 380–82; 1870–1913: Hobson, 'Military-extraction Gap and the Wary Titan', p. 501; 1920–75: Flora *et al.*, *loc. cit.*; 1981–97: SIPRI. Italy: 1862–1973: Flora *et al.*, *op. cit.* vol. i, pp. 402 ff.; 1981–97: SIPRI. Russia: 1885–1913; Hobson, 'Military-extraction Gap and the Wary Titan', p. 501; 1933–8: Nove, *Economic History*, p. 230; 1940–1945: Harrison, 'Overview', p. 21; 1985–91: IISS, *Military Balance*; 1992–7: SIPRI. GDP/GNP/NNP: UK: 1850–70: Mitchell, *European Historical Statistics*, p. 408; 1870–1948: Feinstein, *National Income, Expenditure and Output*, statistical tables, table 3; 1848–1998: ONS. US: 1850–1958: Mitchell, *International Historical Statistics: The Americas*, pp. 761–74; 1959–98: Federal Reserve Bank of St Louis. Germany: 1870–1938: Hoffmann, Grumbach and Hesse, *Wachstum*; 1950–60: Rytlewski (ed.), Bundesrepublik in

Ferguson, *The Cash Nexus*, 2001

Spurious knowledgeability: unnecessary graphical complexity

There are many other ways of intimidating readers, some of which lend themselves better than footnotes to charts for live presentations. A suitably cowed audience is unlikely to ask awkward questions when the lights go up for fear of exposing humiliating ignorance or stupidity. Many such techniques work by making audiences ask themselves if an otherwise inexplicably complex chart doesn't just demonstrate that they are too dim to understand it.

In this example from the *McKinsey Quarterly* the eye is caught by the unusual sequence of axis values above the dots. Puzzlement deepens when it becomes clear that the product sectors are already ranked by expected value of imports into the USA from low-cost countries in 2015 and that few, if any, spectacular changes of rank are expected between 2002 and 2015. Another distraction is the fact that the dots seem to go down to about rank 35 and start at slightly below rank 1 although only 13 categories are displayed. What are we supposed to make of the 20-25 apparently missing industrial sectors? Why use ranks and not

Projected US imports from low-cost countries in 2nd-wave skill-intensive sectors,[1] 2015

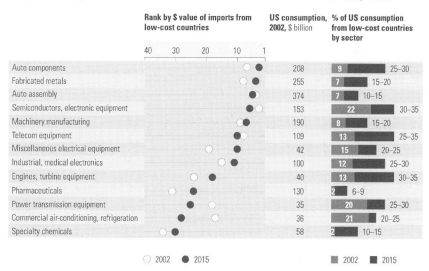

[1]Consumption is assumed to grow in line with GDP.

Source: Bureau of Economic Analysis (US Department of Commerce), Centre for Monitoring Indian Economy (CMIE), Economist Intelligence Unit, McKinsey analysis.

Balasubramanian & Padhi, *The next wave in US offshoring*, 2005

absolute values? As the questions pile up, so audience morale and self-confidence slumps.

Using the data from the bars on the right, the answer may be that the authors faced a graphical problem in depicting import values ranging from about US$1 billion (bottom line: specialty chemicals in 2002) to about US$60 billion (top line: auto components in 2015) on one chart. In fact, a 60:1 value spread is not all that challenging, especially as the chart shows no signs of being cramped for space. Slightly unusually for a McKinsey chart there is no action title, which makes it all the harder to assign a PDQ value. Even after consulting the surrounding text, and assuming that such an august source of graphics is not merely incompetent, it is difficult to see any other reason than managing reader expectations for resorting to such a counter-intuitively orientated and eccentrically dequantified axis definition.

Graphics for graphics' sake

Many graphical deception techniques seem to serve no useful purpose from either the victims' or the perpetrators' point of view and can only be explained under the broad heading of 'graphics for graphics' sake'. What they have in common is that the reader or audience is mainly aware of the graphic technique rather than the message. In this they differ only in one important respect from over-enthusiastically applied make-up. While both obscure the main features, the message in the latter is traditionally considered to be fairly clear.

Both spring from fear of being visually boring or downright ugly. Carried to excess both often reveal rather than hide severe personality disorders.

Let baubles dominate data

Turning bars and columns into visually more exciting icons produces, as we have seen, so many problems that it's seldom worth the effort, unless some really serious deceit is afoot. If all you want to do is obscure and confuse, or just catch readers' eyes without being terribly concerned about whether the message gets lost in the process, it's usually enough to put a perfectly ordinary bar, line or pie chart on a colourful or, even better, distorting background or to adorn it with glittering iconic baubles.

Interest rate changes

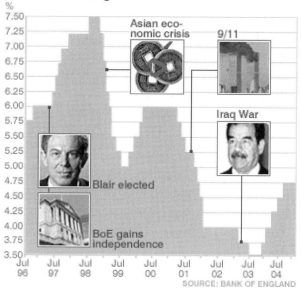

BBC, 2005

In this BBC example from early 2005, the pictures give overriding emphasis to four events: New Labour winning the 1997 election and the subsequent freeing of Bank of England interest policy from government control; the economic crisis in Asia in 1998; the terrorist attacks on the Twin Towers in 2001 and the Second Gulf War in 2003. Yet as the chart itself shows on closer inspection, none of these events, with the possible exception of the first, seems to have had much effect on interest rates. The Asian crisis, 9/11 and the Iraq war all occurred half way down a steep decline in interest rates. Leaving aside the fact that the value axis has been broken and halved in height to dramatise changes, itself a fairly eye-catching distortion technique, the pictures only add information for those who don't know what Messrs Blair and Saddam Hussein and the façade of the Bank of England and the World Trade Centre look like, or are unaware that some Asian coins have holes in them. Displaying interest rates to such a group borders on supererogation.

Yet we should not smirk too soon, as two strong subliminal messages are in fact generated by the use of these icons. First, that three out of the four most powerful influences on the UK economy since 1996 were events over which no British government, let alone parliament, had the

175

slightest control. On second thoughts, it could even be four out of four. Second, that interest rates and thus economic well-being, are exclusively determined by dramatic events rather than by more continuous factors like, for example, growing British and American balance of payments deficits. The fact that the first proposition is probably as true as the second is false simply serves to demonstrate the power of this entertaining graphical technique. Being themselves event-driven, news services like the BBC may not even notice the distortion.

Let background dominate data

All graphics by definition employ metaphors, but some are more meta-phorical than others. Sometimes the metaphor escapes from its graphical cage, takes on a life of its own and provides exciting deception opportunities. Take for example the strange case of the the vacuum cleaner and Britain's leading contribution to international cuisine: Tesco. Only a spoil sport could begrudge *The Times* supporting the headline 'Tesco cleans up' with a picture of a gigantic vacuum cleaner sucking in £1 coins. With a retail turnover of around £30 billion and over a fifth of the UK food market, Tesco puts other British shopping names in the shade.

But then unfortunately the metaphorical vacuum cleaner really starts to suck. Foreign turnover of £8 billion is not to be sniffed at but represents such a tiny market share in, say, the Far East, that to portray Tesco as devouring everything eastnortheast of the Himalayas is an exaggeration that The Times should have resisted. It also distracts from the debate about Tesco's dominant market position in the UK that probably prompted the chart in the first place. To put it another way, the vacuum cleaner is sucking at the wrong, eastern, end of the Eurasian landmass.

Having sacrificed content to layout with the vacuum cleaner, it was but a step to the highly misleading suggestion that '£1 out of every £8 spent in the UK goes to Tesco'. In fact on Treasury figures for 2004 Tesco 'only' accounts for about £1 in £25 of private consumption (excluding business and government). You only get to the 1 in 8 figure by narrowing the focus to sales in retail shops, which admittedly is still highly impressive and, possibly, worrying.

These first two examples of graphics for graphics' sake show how we can use exaggerated visual treatments to add tendentious, or simply

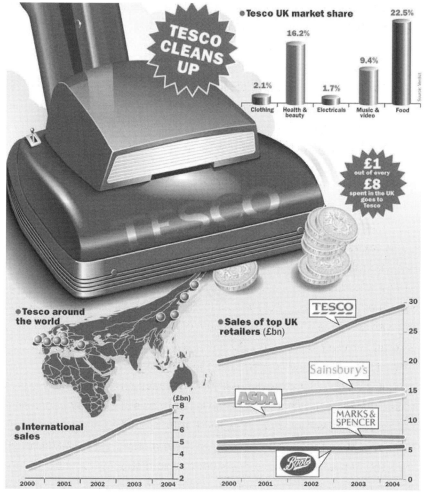

The Times, 2005

erroneous, messages. Their strength and interest lies in their STD values. It seems so puritanical and small minded to want to deny newspaper readers a little dash of colour in their drab and uneventful lives that some powerful distortions slip through with a jovial laugh.

Chart junk

In their purest form, graphics for graphics' sake lead to chart junk. Although pictures of Saddam Hussein and Tony Blair or vacuum cleaners distract and irritate they retain a link, however tenuous, to the message of the chart. They are still in some sense representational, or at least

Impressionist or Surrealist, in the way that they toy with reality. But graphics, like painting, move on and have long employed techniques so abstract as to gladden the heart of a Modigliani or Pollock.

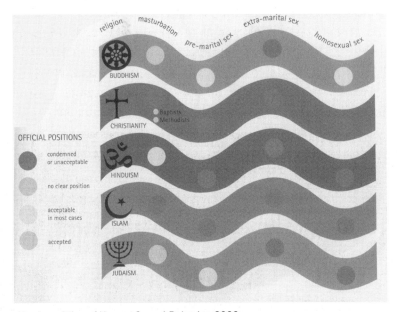

Mackay, *Atlas of Human Sexual Behavior*, 2000

This particular example wears its history on its face. It clearly started life as a simple matrix with religions down one side, various sexual practices across the top and marks at their intersection ranging from A for 'Aaargh, you beast' to Z for 'Zounds fun'. 'Official Positions' (geddit?) probably came next, followed by coloured dots to replace A to Z, leading to a harmless, if boring, chart looking a bit like a modern packet of pills. The enlivenment seems to have started with jauntily misaligning the rows. This however threatened to insert an unwarranted degree of ecumenical ambiguity into the display, which in turn required reinforcing the rows with entirely appropriate serpents. The key point of the chart, that only Islam and Christianity entirely fulfil Mencken's definition of Puritanism[1], gets lost in the junk. It is difficult to say whether this is deliberate or not, always the hallmark of a great graphic.

[1] 'Puritanism – The haunting fear that someone, somewhere, may be happy.' Henry L. Mencken, *A Mencken Chrestomathy*, Sententiae, 1949.

Irrelevant, inconsistent or simply wrong icons

These two examples return to naturalistic art and even to vaguely relevant icons, but with an additional twist. The snag about icons from our point of view is that if they are prominent enough to distort or distract, the reader or viewer is all too likely to notice if we have got them wrong. Even quite small details can betray and diminish the plausibilty of the whole chart. On the other hand, such niggling errors may simply increase the distraction, filling the reader with a warm glow of satisfaction at being more knowledgeable than the author.

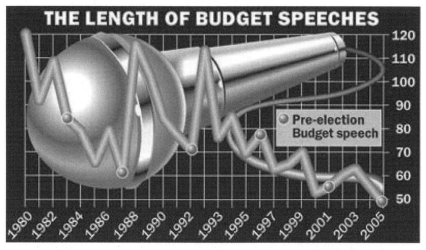

The Times, 2005

The length of the speech that the Chancellor of the Exchequer uses to introduce his annual budget to the House of Commons varies quite surprisingly, though there has been a general tendency in the last 25 years for speeches to become shorter. Curiously, budget speeches before general elections tend to be shorter than those earlier in the life of a parliament. Other traditional features of this annual event have been known to include a modest glass of whisky to lubricate the Chancellor's delivery and many mourned the retirement of the battered, not to say tatty, despatch box he displayed to television viewers on his way to the House. This may not be hi-finance, but for aficionados of the politics of economics it is at least entertaining.

However the one aid to delivery the Chancellor does not use is a hand-held microphone, even if some budgets of modern times would not have suffered from being set to music and delivered as rap titles. The result is a neatly irrelevant background with the added twist of distracting us with questions about the meaning of the icon.

This example from the German Federal Statistical Office is more subtle and even more distracting. The Federal Statistical Office generally views attempts to add sex appeal to its graphics as a cardinal would view attempts to add lap dancing to the Liturgy. But for some reason it recently unbent to the extent of deigning to put a euro coin behind a pie chart about Germany's share of the the economy of Euroland. The result is not exactly a great chart but it was a gallant attempt to move with the graphical times. Unfortunately the statisticians chose the wrong, specifically German, side of the coin, sug-

German % share of the nominal GDP of Euroland, 2001

Anteil des deutschen nominalen Bruttoinlandsprodukts am Bruttoinlandsprodukt der Eurozone 2001[1]

Deutschland
30%
2 064 Mrd. Euro

gesting to all those who have ever spared a glance for the currency that the 30% is a share of something wholly German. Using the other side, which is common to all euro coins wherever they are minted, would have correctly suggested a German share of the Euroland economy.

Some graphical inaccuracy is so minor that it scarcely seems worth mentioning, were it not for the disproportionate distraction it can cause. In this case from the *Atlas of Human Sexuality* the inaccuracy is compounded by the fact that no comparison of any sort is shown, which suggests that the entire chart is unnecessary. But the error distracts from

A study of infertile men in Italy showed that their sperm count was at its highest at around 5 o'clock in the afternoon.

Mackay, *Atlas of Human Sexual Behavior,* 2000

this too and the result is a gaudy monument to graphics for graphics' sake which will probably go unnoticed by all but the most fastidious.

Does this sort of graphic inaccuracy make much difference to our powers of distortion? Probably not very much, but it's a good illustration of the power of minor mistakes to distract, which deserves our grateful emulation.

In 1926, after seven years' delay on the slipway at Swan Hunter caused by the indecently sudden end of the First World War, a new destroyer was launched and given the stirring name, redolent of all that is most glorious in the British naval tradition, of *HMS Whitehall*. It is said that the Admirality gave her captain dire warning that using any of the more obvious symbols associated with this name in his design of the ship's crest would make it not only his first, but also his last, independent command. So she sailed, in this sense if no other, unmarked by any suggestion of the thin, red, linen tape, which threaded through a hole in the top left corner of official documents in lieu of a paper clip or even tied parcelwise around larger bundles of files, had become a byword for bureaucratic incompetence. There are still lawyers' offices with a similar approach to data storage, though the tape used is often white.

Office technology changes slowly – it is, for example, but ten years since French banknotes were still printed with an empty white space for the insertion of pins to hold together bundles of ten – but the arrival of paper clips, lever arch files and the stapler followed a few generations later by electronic data storage led gradually to the disappearance of red tape, even in Whitehall. Yet the name lingers on as the sort of metaphor George Orwell described as having been worn smooth by constant use, like an old coin. Nowadays it lacks the slightest connection to daily experience and so has turned from a resonant metaphor into an exhausted cliché (as has the word 'cliché' itself in most people's minds). As a side effect, 'red tape' has become a graphical no-go area, as any icon will inevitably either be unknown to the under-65s or simply wrong.

The Conservative Party's graphic designer transformed pale red linen tape into a roll of bright red, glossy self-adhesive tape in a transparent plastic dispenser. In the 2005 election campaign, this icon became the leitmotiv of publicity material for policy areas ranging from reduction of

Conservative Policy Unit, *The Drivers of Regulation*, 2004

government waste to defending plucky Britons from the depredations of 'Brussels'. A year later it still lurked mournfully on their website. How many ministries actually use, or have ever used, red sticky tape? Come to that, when did a 'faceless bureaucrat' last don or doff a bowler hat? This is the same school of graphics that would illustrate the phrase 'the Devil to pay' with a gentleman with horns, scarlet tights and an outstretched palm* or the Privy Council with ... but you get the idea.

Of course, some ill-judged uses of icons are even more obvious and the next example has some claims to be considered the worst graphic ever produced in earnest. The facts were quite simple, indeed so simple that a chart was probably not needed at all.

---

* As Orwell points out, the 'Devil' in question is the longest seam or joint on a wooden deck, which would normally be sealed, or 'paid', with tar or pitch. Hence, as a metaphor for inadequate resources: 'the Devil to pay (and only half a pail of pitch)'.

The Decline of the English Language, George Orwell

A Danish study found a link between the amount of alcohol drunk by aspiring mothers and the probability of their becoming pregnant within six months of ceasing to use contraception. Some 64% of those who drank fewer than five glasses a week became pregnant compared to only 55% of those drinking more than ten glasses. For the obsessively graphicophilic this relationship could have been shown in a perfectly respectable, if abominably boring, column chart (left) and, if necessary, obscured in an equally perfectly disreputable double pie chart (right). Note how the pies have been rotated to destroy the last chance of comparing the size of segments.

Proportion of Danish women becoming pregnant within 6 months of ceasing contraception

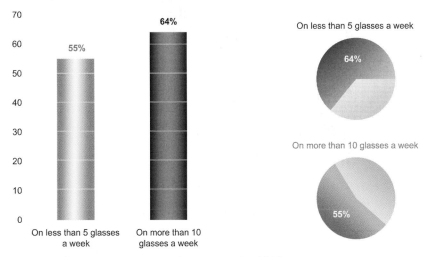

Data from: Mackay, *Atlas of Human Sexual Behavior*, 2000

If it is unnecessary to show readers what 55% and 64% look like, upon which reasonable people may differ, how much more unnecessary was it to show what five and ten glasses look like, on which reasonable people may not? Note too how the lines of glasses have not been lined up either on the left, the right or even the middle, making the comparison of 5 and 10 glasses unusually complex for the graphically challenged reader.

On the other hand, in case all else fails, wilful visualisation of the wrong variable could be a technique to keep in reserve, as it is guaranteed to produce helpless hilarity or even foam flecked rage in virtually any reader.

Outraged convention

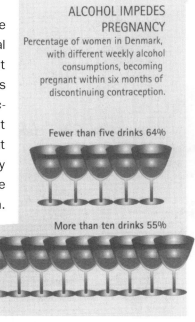

ALCOHOL IMPEDES PREGNANCY
Percentage of women in Denmark, with different weekly alcohol consumptions, becoming pregnant within six months of discontinuing contraception.

Fewer than five drinks 64%

More than ten drinks 55%

Mackay, *Atlas of Human Sexual Behavior*, 2000

This book does not pretend to share very many of the concerns of traditional tomes on graphics. But in one respect there is complete agreement. Readers and viewers have very definite expectations about the orientation, layout and even colour of chart elements that can be turned to our advantage in many different ways but which should only be disappointed for a good, or bad, reason. Changing an established orientation can often amount to a serious political statement in its own right.

Australians have long laboured under the subliminal suggestion that their country is of less importance than its huge size and intellectual and sporting achievements would suggest because it is traditionally depicted as being at bottom right of most maps of the world; the corner generally reserved for the least important information on a chart. There is not the slightest reason in cartography, astronomy or graphics for showing North as the top of the page, except for tradition. Reverse the tradition as here, and the world indeed suddenly looks very different when Western Europe is relegated to the Australian position.

In this example about looming bankruptcy of the pensions system, the German news magazine *Der Spiegel* was trying to make the point that not only had the expenditure of the German pensions system long exceeded its income, but also that the system's forecasts of income and expenditure had been disconcertingly inaccurate for four years. Unfortunately, the inaccuracy of the forecasts in some years was not quite severe enough to be easy to show graphically and so the chart uses simple numerical labels to display the difference between forecast and actual. So far, so barely adequate, since visual comparisons between two series of columns placed one above the other are almost impossible.

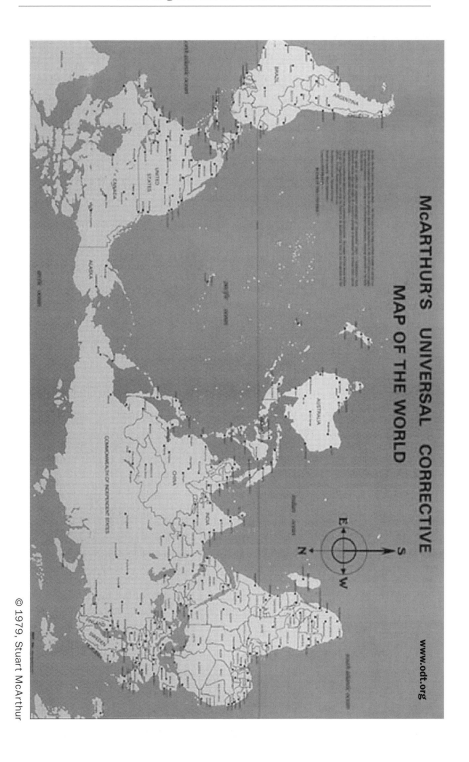

© 1979, Stuart McArthur

Contributions to (top) **and payments by** (bottom) **the German pension system** €10[9]

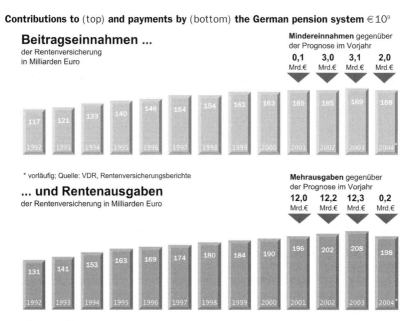

Beitragseinnahmen ...
der Rentenversicherung
in Milliarden Euro

Mindereinnahmen gegenüber
der Prognose im Vorjahr

| 0,1 | 3,0 | 3,1 | 2,0 |
| Mrd.€ | Mrd.€ | Mrd.€ | Mrd.€ |

* vorläufig; Quelle: VDR, Rentenversicherungsberichte

... und Rentenausgaben
der Rentenversicherung in Milliarden Euro

Mehrausgaben gegenüber
der Prognose im Vorjahr

| 12,0 | 12,2 | 12,3 | 0,2 |
| Mrd.€ | Mrd.€ | Mrd.€ | Mrd.€ |

Der Spiegel, 2005

But something seems to have awoken the fiend that sleeps in every graphic designer and the little downward-pointing triangles appeared. In the top graph, this corresponds to traditional expectations that shortfalls in some sense point downwards. But in the bottom half the same orientation represents increase.

A more quantitative version of orientation reversal is provided by this example from the *Wall Street Journal*. This chart is gratifyingly difficult to understand until the realisation dawns that it is simply upside down. Loss of anything is an essentially downwards concept and putting Italian newspapers, which have experienced the sharpest decline in circulation, on the left, the position usually reserved for winners, compounds the problem.

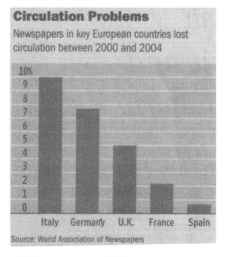

Circulation Problems

Newspapers in key European countries lost circulation between 2000 and 2004

Source: World Association of Newspapers

Wall Street Journal, 2005

Neither of the last two examples causes more than a passing malaise in readers and viewers and both are presumably inadvertent. But a comparison with the inverted map of the world makes the point that unfamiliar graphical orientation is always interpreted initially as deliberate and so as containing meaning. The conscious deceiver is therefore well advised only to allow disorientation in a worthy cause.

The list of the most frequently enforced graphical conventions is surprisingly long and brutally arbitrary. But with occasional exceptions it has remained remarkably stable over long periods of time, even decades.

Time/causation: ➜, *flows from past/independent variables to* ➜ *future/dependent variables. Past:* ⬅

Improvements: ↗. *Best:* ⊞, *worst:* ⊞

Deterioration/failure *to meet a target* ↓, ↘

Numbers, *positive:* ↑, *negative:* ↓

Compass: *North:* ↑, *South:* ↓, *East:* ➜ *& West:* ⬅

Darker *is more important/recent; in 3D, further away*

Symmetry *is better than asymmetry*

Bigger *beats smaller.*

Red: *no/dangerous/bad/important/American/communist/stop!*

Blue: *male/naval/European/British/Conservative/gay* (slav)

Purple: *religious/catholic*

Pink: *female/gay/happy*

Black: *bad/true*

Green: *good/yes/go/muslim*

Even stranger are the colour conventions, which make those quaint Victorian manuals about the language of flowers seem rational. But whatever we may think of these conventions, they exist and deviation from them will nearly always provoke a wondering frown of incomprehension.

Fake graphics and chart thingies

As the reader will have noticed, this book is always at pains to treat both data, graphics and the sources of the most unsatisfactory charts with unflinching respect and gravity. This is not just obsequiousness. Disrespect is usually but a prelude to lowering one's guard, be it towards a technique, an idea or an opponent. So when the National Blood Service nervously giggles 'Here's one of those chart thingies to show you the national distribution of the various blood groups', there is an understandable tendency to expect the worst. To be fair, the NBS does not disappoint.

Distribution of UK blood types in % of population

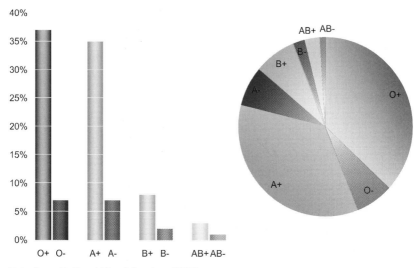

Data from: National Blood Service, 2006

The geographic distribution of blood groups in England still shows traces of the arrival of successive waves of slavering Scandinavians during the Dark Ages and of their Norman descendants in the mid-eleventh century. Indeed worldwide variations in the distribution of blood groups are a sort of Debrett's, Almanac de Gotha or similar Turf Guide to the entire human race. From a more personal point of view, if you are considering donating blood, it's interesting to know what proportion of the population stands to gain from your generosity or, to look at it another way, whether you are potentially helping a group so small, for example AB negative, that it might have difficulty finding adequate supplies of matching blood without your help. Whatever your interest, the distribution of UK blood groups can reasonably be summarised in a bar or pie chart. The length of the bars and the size of the pie slices are, fairly obviously, proportional to share of population with that blood group.

But it was not so obvious to the National Blood Service, which chose instead to make the lengths of the bars proportional to the width of the name of the blood group. What turns this web page from a mere missed graphical opportunity (with ham-fisted typography) into part of an ingenious attack on the reputation of the National Blood Service is that initial giggle announcing a chart. Would you be happy to have the publisher of this 'chart thingy' take a needle to your veins and open the tap?

Moiré or other distracting distorting hatching

It has been said with only mild hyperbole that the history of scientific advance is actually the history of improving power and precision of scientific instruments. Galileo would never have been left muttering 'E puer si muove' (But it [the planet Earth] does move) if advances in grinding and polishing optical lenses had not produced telescopes capable of raising awkward questions about the reliability of papal cosmology.

Most changes in graphical style have been based upon improvements in printing and copying technology, which have broadened the choice of graphical devices available for deception. Luckily the old techniques linger on, long after the technology that necessitated them has been abandoned.

All about your blood types

Here's one of those chart thingies to show you the national distribution of all the various blood groups.

Click on your blood type below for more information.

Frequency of major blood groups in the UK population

| Total Blood Type O | 64% |

| Total Blood Type A | 42% |

| Total Blood Type B | 10% |

Total Blood Type AB	4%
Total	Rh (D)	Pos	83%
Total	Rh (D)	Neg	17%

National Blood Service, 2006

For many years photocopiers could only reproduce the sort of thinnish lines required by printed text. Large homogenous areas of any shade, particularly black, came out with elaborate gradations of shading or none at all. Repetitious patterns of fine, closely spaced lines or shading of any type were completely beyond them, as they are for some fax machines to this day. This was even true of the early colour photocopiers. The launch of colour copiers that could deal with such patterns led to the pulping of most of the world's banknotes, even of the almighty but easily copiable dollar, and the introduction of notes with surface

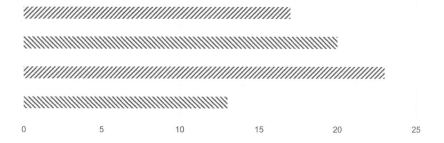

texture, holograms, watermarks and variably embedded metallic strips. Even so, the surprisingly high quality of the forgeries using just a decent scanner and printer, in some sense the ultimate graphical deceit, should be an inspiration to us all.

The limitations of the first photocopiers and of most predecessor printing technologies left its mark on the way we shade the bars and columns in charts. Most of these shadings are still available in Excel to this day and are reminiscent of the hand-cranked rotation facility on early radars to keep the aerial turning in the event of electrical power failure. They resemble, and may even have inspired, the worst excesses of Op art in the 1970s and are almost guaranteed to produce headache and even cortical seizures in their viewers.

Preference for infantry or for overseas combat service as related to worry about battle injury
(Infantrymen in the United States, 3 months to 3 years of service, April 1944)

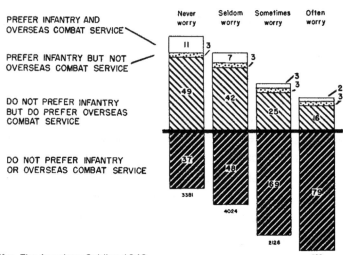

Stouffer, *The American Soldier*, 1949

The real life example from a 1950s book on the American Soldier illustrates the unsettling effect of diagonal hatching on the staidest graphic and is a good introduction to the use of optical illusions in general to confuse and repel readers.

Excel kindly puts a very much larger range of cortical solvents at our disposal. There may even be residual legitimate markets for these shadings and hatchings, just as there are for buggy whips, musket flints and crinoline hoops. But it is reassuring to know that if we ever need to give anyone an instant migraine the means are at hand.

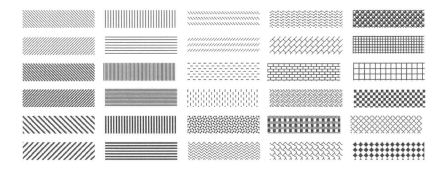

Legends

If you sincerely hate your readers there is a variety of other graphical techniques to make them wish they had never opened your book, without inflicting massive neurological dysfunction on the innocent.

One of the most elegant is using legends rather than labelling graphical elements directly. Despite their name, which evokes the inspiring but mendacious, legends often present themselves successfully as kindly assistants who can help the dullest reader discern the meaning of the chart.

Legends run from the tiny but perfectly formed to the seriously evil.

Minor examples like this chart from a learned paper on molecular biology illustrate the problem in its purest and usually most innocent form. There is no pressing graphical reason to abbreviate the names of the long bones of the mouse skeletons, for which there is plenty of space below each pair of columns. There is also space enough to identify the paired columns with adjacent labels reading 'wild-type' and 'mutant' instead of 'wt' and 'm'. Placing the legends as far as possible from their

**No statistically significant difference in the length of the long bones
of newborn limbs between wild-type and mutant littermates ...**
Skeletal length in mm (n = 5), error bars represent ± standard errors of the mean

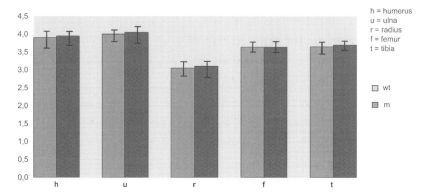

Ko et al. *Matrilin-3 is Dispensable for Mouse Skeletal Growth and Development*, 2004

reference is less quantitative communication than compulsory callis-
thenics for the eye and neck muscles. Unless you remember that using
legends is the default option on most graphics programmes it's often
difficult to imagine how the subject even came up.

More complex charts can achieve awesome levels of communicative
dysfunction with the help of legends. As with some other charts in
Ferguson's *The Cash Nexus* there is a slight ambiguity of purpose in this

THE COMMONS AND THE CASTLE

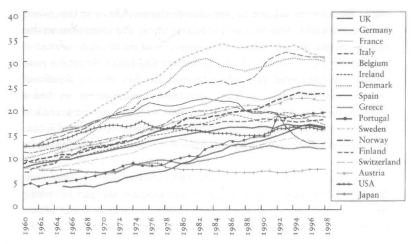

Government employment as a percentage of total employment

Ferguson, *The Cash Nexus*, 2001

example. The accompanying text states 'Figure 7 shows that this growth [of the government share of total employment] has since continued in more or less every developed country and has been substantially reversed in only one', but the feeling persists that it could have been more explicitly reflexive: 'Get me!', perhaps. Whatever the motivation, it needs a pick and a miner's lamp to find the curve for the exceptional developed country and a powerful magnifying glass to identify it by reference to the legend as belonging to the UK, rather than Germany or Spain.

Legends, in their own or anyone else's lifetime, belong in Greek myths, Viking sagas and Le Carré novels, not in charts. Unless of course deterrence is your aim, in which case they are almost indispensable.

The Labour Party's manifesto for the 1983 election was memorably described by Gerald Kaufman as 'the longest suicide note in history' but the Green Party in 2005 almost set theirs to music with the help of multi-coloured legends requiring highlighted footnotes for correct use. After the outrageously populist title 'Savings curve for CO_2 abatement options' the most striking element of Figure 1 is the instruction at the bottom on how to read the legend, or key as it is called here. Quite apart from the redundant, and slightly confusing, 'starting', deleting both the

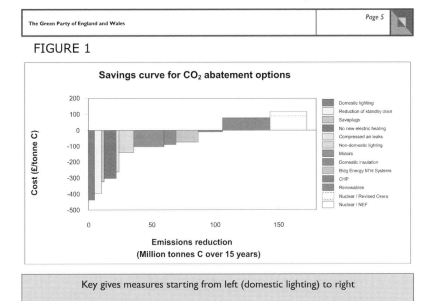

FIGURE 1

Toke D. & Taylor S., *The Alternative Energy Report*, The Green Party of England and Wales, 2006

legend and the explanation would have created enough space to label the columns directly and comfortably. Clearly the Greens are not a party for the faint-hearted. It is just possible still to sympathise with a political party that requires its activists to have a firm grasp of the scientific abbreviations for the chemical elements, but to extend this requirement to potential voters borders on the reckless.

Other repellent or illusion-creating charts

If the point of quantitative visualisation is to summon spatial relationships to the aid of the data, optical illusions in charts come close to Original Sin. Unfortunately most illusions are so well known or so awkward to achieve that STDs are impossibly high. The only major exception is the Müller-Lyer illusion that has been granted new vigour by the increased availability of 3D graphics software packages.

The Müller-Lyer, or arrowhead, illusion is the most powerful optical illusion known to neurophysiology. In its simplest form the Müller-Lyer can transform the apparent lengths of any pair of lines by adding inward or outward facing 'arrowheads' to their extremities. In this crude example the red lines are of equal length but the line on the right appears to be longer.

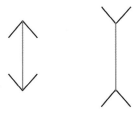

There are various explanations for this effect, but it is probably caused at a surprisingly early stage in human visual processing that interprets the convergence of three straight lines as indicating a 3D corner. Further up the processing chain this is reinterpreted as meaning that the lines on the left are an outside, and therefore nearer corner, while those on the right are an inside, and therefore more distant corner, assuming that the outermost bits of the arrows are all equidistant from us. Our logical monkey-type brains then reason backwards from the fact that the red lines subtend an equal angle on our retinas to the conclusion that the more distant line must therefore 'in reality' be longer. QED, or alternatively: neat, but not gaudy.

With a little care, Müller-Lyer illusions can even be worked into chart backgrounds of the sobriety of the Vietnam War Memorial in Washington.

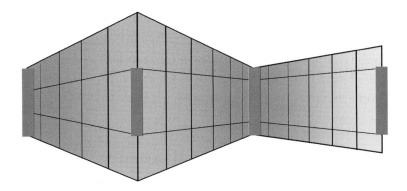

The illusion has been reinforced here by shading and different line thicknesses to exaggerate the feeling of 3D depth. It is powerful enough to have strong men reaching for pencils and rulers and drawing construction lines on their handouts to see if they can trust their own eyes. Only some tribes of Bushmen, whose culture knows few straight lines, are rumoured to be partially immune, as their brains don't automatically assume that three converging, straight lines are the corner of a solid object. In the Kalahari apparently, an arrow is an arrow, is an arrow and not the corner of a modern building.

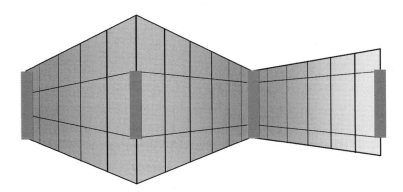

For those who still find it hard to believe, here the picture is repeated with straight lines connecting the tops and bottoms of the red columns. Susceptibility to the Müller-Lyer illusion varies but some readers may even detect a slight bending of the straight lines, particularly near the third column from the left, as their brains frantically try out different hypotheses to make sense of an impossible visual image.

A domesticated version of the Müller-Lyer, in which depth is suggested less by the angles of pediments and pedestals than by other depth clues, is almost unavoidable in any 3D chart with grid lines unless a strictly isometric version is selected. Isometric representation has the advantage of maintaining the same scale throughout (unlike this example in which scale becomes dramatically smaller with increasing virtual distance from the viewer) and can be useful for, say, shopfitters who might want to measure an unlabelled distance on the blueprint accurately. But it obviously leads to unaesthetic, and possibly confusing, angular distortion.

Seckel, *Incredible Visual Illusion*, 2003

The gorget illusion is old and takes its name from an article of military clothing, a half-moon shaped fragment of vestigial armour worn to protect the throat, not seen for a couple of centuries. In this case the old ladies seem to decrease in size from left to right. If you're not very susceptible to this illusion it helps to focus on an adjacent pair and let your mind go limp; the illusion will soon appear in spite of the fact that these triplets are absolutely identical in size and shape.

Also known as the Wundt illusion, it seems to depend on our inability to judge the real diametre of segments of a circle, when they are stacked close to one another. All three of the circle segment shown here have precisely the same diametre although, as with the triplets, the circles from which they are cut seem to get smaller from left to right. This is the Tolansky illusion.

It's not easy to think up a practical graphical application for the gorget illusion, still less to find a real life example. However the *Harmsworth Atlas* published in about 1908 neglected no aspect of the British Empire and its dominance of most of the world's commerce. Even the banana trade, and its exciting growth of 2.7 times between 1900 and 1905, which would put today's Chinese and Indian economies in the shade, did not go unrecorded or uncelebrated. At first, this seems to be a real live sighting. But sadly,

GROWTH OF BANANA TRADE OF UNITED KINGDOM.

1905

1900

£548,956   £1,498,084

Harmsworth Atlas, 1907

although the banana on the right looks distinctly less than 2.7 times the size of the banana on the left, this is largely because it really is only twice (1.92 times) as big.

Quite why Harmsworth decided to downplay this spectacular British success story remains obscure but the means chosen was run of the mill scale fraud, with a bit of length/area/volume ambiguity thrown in for good measure, rather than the more elegant gorget illusion. Pity!

Unnecessary icons

It seems fitting to close this chapter on graphics for graphics' sake with the corniest, and from the average reader's point of view most insulting, technique of all. The examples are all the more striking for being drawn from the website of the US Bureau of the Census, which normally stands in the same relationship to American numbers as Henry James or Gore Vidal to American letters: as the acme of authoritative elegance.

The topic was illegitimacy rates, or more precisely the role which shotguns and other quintessentially American artefacts may have played in their limitation. The figures, based on careful comparison of the dates on a very large number of marriage and birth certificates, showed that whereas in the 1930s only one in six first births to young women were conceived before marriage, the proportion had risen to half by the early 1990s, assuming of course that gestation periods had not changed significantly over the period surveyed.

One in Six First Births to Women 15 to 29 Years Old in 1930–34 were Conceived before First Marriage

Bachu, *Timing of First Births: 1930–34 to 1990–94*, U.S. Bureau of the Census, 1998

The findings are difficult to interpret. One of the goals of the institution of marriage, so bluntly expressed in the *Book of Common Prayer* as the 'avoidance of fornication', is clearly not being met. On the other hand, and on the principle of 'better late than never', there's a lot of life in the institution yet. Indeed not only Americans indulge in it again and again.

From a graphical point of view these are charts to treasure, though not all of us will have the necessary thickness of hide to emulate them.

One in Two First Births to Women 15 to 29 Years Old in 1990–94 were Conceived before First Marriage

Bachu, *Timing of First Births: 1930–34 to 1990–94*, U.S. Bureau of the Census, 1998

Counter-message ink

The graphic instinct behind this example is quite sound. Given two and a half pairs of factettes about American opinions on cloning, the simplest form of display is indeed just to make a list like this. But, of course, if everything else in the book has received some sort of graphical treatment, it's difficult to avoid it here. Luckily the book has already established its own unmistakeable iconic norms and so the mysterious ball and oval is more or less recognisable from earlier chapters as ushering in the results of opinion polls. Even the micro-phone, not much used by pollsters, could just about slip by as a metaphor for recording *vox populi*. But the masterstroke, dazzling in its ambiguity of intent, which places this chart only a little lower than the angels, is the microphone cable.

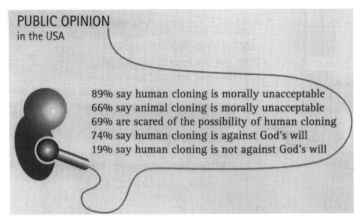

PUBLIC OPINION
in the USA

89% say human cloning is morally unacceptable
66% say animal cloning is morally unacceptable
69% are scared of the possibility of human cloning
74% say human cloning is against God's will
19% say human cloning is not against God's will

Mackay, *Atlas of Human Sexual Behavior*, 2000

Most readers will already have begun developing and discarding hypotheses long before they get to the text. Is this just a happy accident which the proofreader failed to spot, or is even posing the question evidence of such an irrevocably polluted mind that it would better be left unasked? Or then again, is this perhaps the work of some sly and disgruntled graphic designer out to get his revenge on the unfortunate publisher or did it have the publisher rolling in the aisles too? What's the connection, if any, with South Africa? And as for the proportions, the imagination starts to spin gently out of control, unless of course the suggestion is that some men might be relieved to hand the whole business over to a technician in a cloning lab. Whatever the truth of the matter, one effect is absolutely clear. No one is going to bother with the numbers. This is a technique to treasure and is far too rarely encountered.

Is there such a thing, can there be such a thing, as an all-embracing general theory of graphic deceit? As the last two hundred pages demonstrate, there is a dizzying variety of ways in which graphics can be used to persuade, impress and confuse. This book celebrates the inexhaustible creativity of the human spirit and its dazzling ability – while communicating facts and figures – to escape the bonds of taste, clarity and truth. Given a sincere desire to mislead, modern graphics software and a reasonably powerful PC, there is no practical limit to the number of ways we can deceive without even, quite, lying.

So the best we can realistically hope to achieve is a framework for understanding what can be done to whom, when and how. The relevant techniques of deception and obfuscation have been catalogued and explained here in the hope of inspiring emulation of the best or of achieving immunisation against the worst. Readers must decide for themselves which of these two aims they find the more immediately appealing.

It is not, after all, the facts and figures in life that matter, but the courage one brings to their presentation.

CREDITS

1 Introduction Page 2 top: Mackay J., *The Penguin Atlas of Human Sexual Behavior*, p.63 © J. Mackay 2000; **2** bottom: *The Gettysburg PowerPoint Presentation* www.norvig.com/Gettysburg © Peter Norvig 2006

2 The Power of Graphics Page 19 *Report of the PRESIDENTIAL COMMISSION on the Space Shuttle Challenger Accident*, Washington DC, 6.6.1986, Volume 4, p.651, ref. 2/14–3, p.2 **20** ibid, Volume 4, p.651, ref. 2/14-3, pp.11 & 6 **21** ibid, Chapter V: *The Contributing Cause of the Accident*, p.89 **22** ibid **23** ibid, Volume 5, p. 896, ref. 2/26-2, p.2 **24** ibid, Volume 5, p. 896, ref. 2/26-2, p.1 **25** ibid, Volume 1, Chapter VI: *An Accident Rooted in History*, p.146, figures 6 & 7 **26** Tufte E., *Visual Explanations*, Graphics Press, 1997 © Edward Tufte 1997. Reprinted by permission **28** NASA, www.nasa.gov/pdf/ 2203main_COL_debris_boeing_030123.pdf, p.2, downloaded 29.08.2006 **29** ibid, p.1 **30** Snow, J., *On the Mode of Communication of Cholera*, John Churchill, London, 1855 quoted in: UCLA, School of Public Health, Department of Epidemiology, www.ph.ucla.edu/epi/snow/snowmap1_highres.pdf downloaded 29.08.06 **31** Data only from 'The changing assessment of John Snow's and William Farr's cholera studies' 46(4):225–232 (Soz Praventiv Med) © J.M. Eyler 2001 **32** Data only from Snow, J., *On the Mode of Communication of Cholera*, John Churchill, London, 1855 quoted in: UCLA, School of Public Health, Department of Epidemiology. www.ph.ucla.edu/epi/snow/snowmap1_highres.pdf downloaded 29.08.06

3 Distorting Values: Manipulating the Data Page 36 Data only from Eurostat, http://epp.eurostat.ce.eu.int, downloaded 13.7.2005. © Eurostat **37** *Independent on Sunday*, 29.5.2005 © The Independent 2005 **40** *The Times*, 3.5.2005 © NI Syndication 2005 **42** *International Herald Tribune*, 9.3.2006 © International Herald Tribune 2006 **43** *USA Today*, 27.5.05 © USA Today 2005 **45** Data only from Eurostat, http://epp.eurostat.ce.eu.int, downloaded 13.7.2005. © Eurostat **46** *The Times*, 6.6.2005 © NI Syndication 2005 **47** Data only from UNAIDS, www.unaids.org © UNAIDS 2004 **48** UNAIDS, www.unaids.org © UNAIDS 2004 **49** Data only from Halifax © Halifax 2005 **50** BBC News, http://news.bbc.co.uk/ © BBC 2005, 2006 **51** Data only from Government Actuary's Department, www.gad.gov.uk/Population/index.asp, downloaded 14.7.2005, Crown copyright **52** *The Times*, 20.4.2005 © NI Syndication 2005 **53** *Pre-Budget Report 2004*, HM Treasury, www.hm-treasury.gov.uk/media/98C/55/pbr04_ maindoc1.76.pdf, downloaded 5.7.2005, p.39 Crown copyright **54** Data only from German Federal Ministry of Finance, Berlin, 2005 © Bundesminister der Finanzen **55** *Der Spiegel* 33/2005, p.53 © Spiegel 2005 **57** Smith D. with Sandberg K.I., Baev P., Hauge W. and the International Peace Research Institute, Oslo, *The State of War and Peace Atlas* p.29, Penguin Books 1997. Text © Dan Smith 1997. Maps and Graphics © Myriad Editions Limited 1997

4 Distorting Values: Graphical Manipulation Page 60 Data only from Smith D., *State of the World Atlas*, p.32, Penguin © D. Smith 2003 **61** Smith D., *State of the World Atlas*, p.32, Penguin © D. Smith 2003 **62** top: Smith D., *State of the World Atlas*, p.113, Penguin © D. Smith 2003 **62** bottom: *The Harmsworth Atlas and Gazetteer of the World*, London ca. 1907 **63** *The Harmsworth Atlas and Gazetteer of the World*, London ca. 1907 **64** ibid **65** Data only from BBC News, http://news.bbc.co.uk/ © BBC 2005, 2006 **66** top: BBC News, http://news.bbc.co.uk/ © BBC 2005, 2006 **66** bottom: data only from BBC News, http://news. bbc.co.uk/ © BBC 2005, 2006 **67** top: data only from Smith D., *State of the World Atlas*, p.103, Penguin © D. Smith 2003 **67** bottom: Smith D., *State of the World Atlas*, p.103, Penguin © D. Smith 2003 **69** top: US Bureau of the Census, Population Division Paper 24 April 1998: *Survey of Income and Program Participation* **69** bottom: *USA Today*, 9.10.01 © USA Today 2001 **70** Data only from *The Times*, 18.4.2005 © NI Syndication 2005 **71** *The Times*, 18.4.2005 © NI Syndication 2005 **72** Data only from Deutsche Bahn AG, http://www.db.de/site/shared/de/dateianhaenge/ presse/infografik/bahnuebergaenge. pdf, download 29.8.2006 © Deutsche Bahn AG 2006 **73** Deutsche Bahn AG, http://www.db.de/site/shared/de/dateianhaenge/presse/infografik/bahnuebergaenge.pdf, download 29.8.2006 © Deutsche Bahn AG 2006 **74** Data only from *The Times*, 4.5.2005 © NI Syndication 2005 **75** *The Times*, 4.5.2005 © NI Syndication 2005 **76** Data only from *The Times*, 17.3.2005 © NI Syndication 2005 **77** *The Times*, 17.3.2005 © NI Syndication 2005 **78** Devine J. & Coleman C., *An Evaluation of the State and County Housing Estimates*, Population Working Paper Series No. 71, Washington DC, April 2003 **81** Mackay J., *The Penguin Atlas of Human Sexual Behavior*, p.76 © J. Mackay 2000 **82** Mackay J., *The Penguin Atlas of Human Sexual Behavior*, p.54 © J. Mackay 2000 **83** *USA Today*, 14.4.98 © USA Today 1998 **84** Smith D., *State of the World Atlas*, p.54, Penguin © D. Smith 2003 **85** Data only from *The Times*, 18.4.2005 © NI Syndication 2005 **86** *The Times*, 18.4.2005 © NI Syndication 2005 **88** top: UNICEF, *The State of the World's Children 2000*, New York/Geneva 2000 © UNICEF **88** bottom: *USA Today*, 17.2.99 © USA Today 1999 **89** *Capital* No.16/2005 © Capital 2005 **90** Data only from a transcript of PowerPoint presentation on 30.7.2003, Prime Minister's Delivery Unit, www.number-10.gov.uk/output/Page4295.asp, slide 50 Crown copyright **91** top: PowerPoint presentation on 30.7.2003, Prime Minister's Delivery Unit, www.number-10.gov.uk/output/Page4295.asp, slide 50 Crown copyright **91** bottom: data only from Reckless M. & Tate J., *The Drivers of Regulation* © Conservative Party 2004 **92** ibid **93** Reckless M. & Tate J., *The Drivers of Regulation* © Conservative Party © Conservative Party **95** top: Mackay J., *The Penguin Atlas of Human Sexual Behavior*, p.52 © J. Mackay 2000 **95** bottom: *The Times*, 30.5.2005 © NI Syndication 2005 **97** PowerPoint presentation on 30.7.2003, Prime Minister's Delivery Unit, www.number-10.gov.uk/output/ Page4295.asp, slide 50 Crown copyright **98** Mackay J., *The Penguin Atlas of Human Sexual Behavior*, p.30 © J. Mackay 2000 **99** *Das deutsche Kolonialbuch*, p.103, ed. Sache H., Wilhelm Andermann Verlag, Berlin, 1926 **100** *The Harmsworth Atlas and Gazetteer of the World*, London ca. 1907 **103** Data only from International Monetary Fund, Report of the Executive Board for the Financial Year Ended April 30, 1998, Washington DC, 1998 **104** International Monetary Fund, *Report of the Executive Board for the Financial Year Ended April 30, 1998*, Washington DC, 1998 **105** *Bartholomew Times Atlas* © Collins Bartholomew Ltd 1975. Reproduced by kind permission of HarperCollins Publishers **106** Ferguson N., *The Cash Nexus*, p.97, Allen Lane 2001, Penguin Books 2002 © Niall Ferguson 2001 **107** Data only from Smith D. with Sandberg K.I., Baev P., Hauge W. and the International Peace Research Institute, Oslo *The State of War and Peace Atlas* p.54 (Penguin Books 1997). Text © Dan Smith 1997. Maps and Graphics © Myriad Editions Limited 1997 **108** Smith D. with Sandberg K.I., Baev P., Hauge W. and the International Peace Research Institute, Oslo *The State of War and Peace Atlas* p.54 (Penguin Books 1997). Text © Dan Smith 1997. Maps and Graphics © Myriad Editions Limited 1997 **109** BBC News, http://news.bbc.co.uk/ © BBC 2005, 2006 **111** *Der Spiegel* 17/2005 © Spiegel 2005 **112** *The Rise and Fall of the Great Powers* © Paul Kennedy 1987

5 Distorting Categories: Data Manipulation Page 114 Education at a Glance, 2004 © OECD **115** *The Times*, 22.4.2005 © NI Syndication 2005 **116** Data only from *The Times*, 22.4.2005 © NI Syndication 2005 **117** *The Times*, 22.4.2005 © NI Syndication 2005 **119** *The Times*, 19.5.2005 © NI Syndication 2005 **122** BCG, Senior Management Survey, Innovation 2005, p.15 © Boston Consulting Group 2005 **124** www.warresisters.org, New York 2005 © War Resisters League **126** Mackay J., *The Penguin Atlas of Human Sexual Behavior*, p.37 © J. Mackay 2000 **127** Mackay J., *The Penguin Atlas of Human Sexual Behavior*, p.77 © J. Mackay 2000 **128** *Wirtschafts Woche* 17/2005, p.28 © Wirtschafts Woche 2005 **129** Seager J., *The Penguin Atlas of Women in the World*, p.93 © J. Seager 2003 **130** UNO, www.un.org/Overview/growth.htm download on 30.8.2006 © United Nations Organisation; Inter-Parliamentary Union. www.ipu.org/wmn-e/suffrage.htm **132** top: *Keine Angst vor Methusalem!* Zu Klampen Verlag, Lippe, p.39 © N. Strange 2006 **132** bottom: Statistisches Jahrbuch der Bundesrepublik Deutschland 1977 and following years © German Federal Statistical Office **133** Data only from Eurostat http://epp. eurostat.ec.europa. eu/portal/page?_pageid=1996,39140985 © Eurostat **134** *The Times*, 23.3.2005 © NI Syndication 2005 **135** National Statistics, www.statistics. gov.uk/rpi Crown copyright **136** ibid **138** *The Times*, 10.3.2006 © NI Syndication 2006

6 Distorting Categories and Time: Graphical manipulation Page 140 The data only for 'Fate of passengers and crew in the Titanic disaster' is licensed under the GNU Free Documentation License. It uses material from the Wikipedia article 'RMS Titanic. **141** Adapted from Dawson, R. J. M. (1995). The "unusual episode" data revisited. *Journal of Statistics Education*, 3(3) and from Friendly M., Extending Mosaic Displays: Marginal, Partial, and Conditional Views of Categorical Data, *Journal of Computational and Graphical Statistics*, 1999, 8:373–395 **142** BCG, *Senior Management Survey*, Innovation 2005, p.7 © Boston Consulting Group **143** top: Data only from BCG, *Senior Management Survey*, Innovation 2005, p.212 © Boston Consulting Group **143** bottom: Ferguson N., *The Cash Nexus*, p.212, Allen Lane 2001, Penguin Books 2002 © Niall Ferguson 2001 **144** Data only from Ferguson N., *The Cash Nexus*, p.212, Allen Lane 2001, Penguin Books 2002 © Niall Ferguson 2001 **145** Smith D., *State of the World Atlas*, p.91, Penguin © D. Smith 2003 **146** Smith D., *State of the World Atlas*, p.81, Penguin © D. Smith 2003 **147** Smith D., *Der Fischer Atlas zur Lage der Welt*, p.24 © D. Smith (Fischer) 1999 **148** left: Data only from Gao P., 'Supplying auto parts to the world', *McKinsey Quarterly*, 2004 Special Edition: China today. www.mckinseyquarterly.com/image/article/flash/chart_suau04_02.swf **148** right: Gao P., 'Supplying auto parts to the world', *McKinsey Quarterly*, 2004 Special Edition: China today, www.mckinseyquarterly. com/image/article/ flash/chart_suau04_02.swf **149** top: Data only from Gao P., 'Supplying auto parts to the world', *McKinsey Quarterly*, 2004 Special Edition: China today, www.mckinseyquarterly.com/image/article/flash/ chart_suau04_02.swf **149** bottom: Mackay J., *The Penguin Atlas of Human Sexual Behavior*, p.23 © J. Mackay 2000 **150** top: Data only from *Handelsblatt*, 18.8.2005, p.6 © Handelsblatt 2005 **150** bottom: *Handelsblatt*, 18.8.2005, p.6 © Handelsblatt 2005 **151** Data only from *The Times*, 6.5.2005 © NI Syndication 2005 **152** *The Times*, 6.5.2005 © NI Syndication 2005 **153** Data only from Bachu A., Population Division Working Paper No.14 © U.S. Bureau of the Census, Washington DC 1996 **154** top: ibid **154** bottom: Bachu A., Population Division Working Paper No.14 © U.S. Bureau of the Census, Washington DC 1996 **155** *The Times*, 15.6.2005 © NI Syndication 2005 and *Die Entwicklung der Leichtathletik-Weltrekorde*, Deutsche Leichtathletik Verband on 1.11.2003 **156** ibid

7 Distorting the Whole Chart: Mismatch Title and Data Page 160 *The Sunday Times*, 29.5.2005 © NI Syndication 2005 **161** BBC News, http://news.bbc.co.uk/ © BBC 2005, 2006 **162** Mackay J., *The Penguin Atlas of Human Sexual Behavior*, p.52 © J. Mackay 2000 **163** *The Times*, 22.3.2005 © NI Syndication 2005 **164** Seager J., *The Penguin Atlas of Women in the World*, p.48 © J. Seager 2003 **165** *Wirtschafts Woche* 17/2005, p.141 © Wirtschafts Woche 2005 **166** BCG, Navigating the Five Currents of Globalisation, Focus, January 2005, p.10 © Boston Consulting Group 2005 **168** *A Nation Online: Entering the Broadband Age* © U.S. Department of Commerce, September 2004 **169** top: Data only from *Focus* 38/2004, 13.9.2004 **169** bottom: *Focus* 38/2004, 13.9.2004 **171** *The Times*, 18.3.2005© NI Syndication 2005 **172** Ferguson N., *The Cash Nexus*, p.97, Allen Lane 2001, Penguin Books 2002 © Niall Ferguson 2001 **173** Balasubramanian R. & Padhi A., *The next wave in US offshoring*, McKinsey Quarterly, 1/2005 © McKinsey 2005 **175** BBC News, http://news.bbc.co.uk/ © BBC 2005, 2006 **177** *The Times*, 13.4.2005 © NI Syndication 2005 **178** Mackay J., *The Penguin Atlas of Human Sexual Behavior*, p.73 © J. Mackay 2000 **179** *The Times*, 17.3.2005 © NI Syndication 2005 **180** top: Statistisches Bundesamt, www_destatis_de_presse_deutsch_pk_2002_schaubilder_bip2001 © German Federal Statistical Office **180** bottom: Mackay J., *The Penguin Atlas of Human Sexual Behavior*, p.46 © J. Mackay 2000 **182** Reckless M. & Tate J., *Reversing the Drivers of Regulation* © Conservative Party 2004 **183** Data only from Mackay J., *The Penguin Atlas of Human Sexual Behavior*, p.47 © J. Mackay 2000 **184** Mackay J., *The Penguin Atlas of Human Sexual Behavior*, p.47 © J. Mackay 2000 **185** McArthur's Universal Corrective Map of the World © 1979 McArthur. Available worldwide from ODT, Inc. (1-800-736-1293; www.odt.org; Fax: 413-549-3503; odtstore@odt.org). Also available in Australia from McArthur Maps, 208 Queens Parade, North Fitzroy, 3068, Australia; 0011-613-9482-1055. stuartmcarthur@hotmail.com **186** top: *Der Spiegel* 14/2005, p.95 © Spiegel 2005 **186** bottom: *Wall Street Journal Europe*, 27–29 May 2005 © World Association of Newspapers 2005 **188** Data only from National Blood Service www.blood.co.uk Crown copyright **189** National Blood Service www.blood.co.uk Crown copyright **190** *The American Soldier: Combat and its Aftermath* (Princeton University Press) © Samuel A. Stouffer 1949. www.pupress.princeton.edu **192** Ko Y. et al, 'Matrilin-3 Is Dispensable for Mouse Skeletal Growth and Development', *Molecular and Cellular Biology*, Feb 04, p.1694 © Molecular and Cellular Biology 2004 **193** Ferguson N. *The Cash Nexus*, p.97, Allen Lane 2001, Penguin Books 2002 © Niall Ferguson 2001 **194** Toke & Taylor, *The Alternative Energy Report*, The Green Party © The Green Party 2006 **196** Seckel A., *Incredible Visual Illusions*, p.50 © A. Seckel 2003 **197** *The Harmsworth Atlas and Gazetteer of the World*, London ca. 1907 **198** Bachu A., Population Division Working Paper No.25 © Population Division, U.S. Bureau of the Census, May 1998 **199** Mackay J., *The Penguin Atlas of Human Sexual Behavior*, p.48 © J. Mackay 2000

INDEX